RECEIVED

JUL 0 7 2022

D0397315

NO LONGER PROPERTY OF
SEATTLE PUBLIC LIBRARY

Natural

Natural

HOW FAITH
IN NATURE'S
GOODNESS LEADS
TO HARMFUL FADS,
UNJUST LAWS,
AND FLAWED
SCIENCE

ALAN LEVINOVITZ

BEACON PRESS
BOSTON

Beacon Press
Boston, Massachusetts
www.beacon.org

Beacon Press books
are published under the auspices of
the Unitarian Universalist Association of Congregations.

© 2020 by Alan Levinovitz
All rights reserved
Printed in the United States of America

23 22 21 20   8 7 6 5 4 3 2 1

This book is printed on acid-free paper that meets the uncoated paper
ANSI/NISO specifications for permanence as revised in 1992.

Text design and composition by Kim Arney

Library of Congress Cataloging number 2019056300
Library of Congress Cataloging-in-Publication Data is on file.

To my family

NATURE: *Since I am the whole that exists, how is it possible for a being like you, so small a portion of myself, to comprehend me? Be content, atoms my children, with seeing a few atoms that surround you, with drinking a few drops of my milk, with vegetating a few moments on my breast, and at last dying without having known your mother and your nurse.*

—VOLTAIRE, *The Philosophical Dictionary*

Contents

Introduction I

PART I **Myth** 11

CHAPTER 1   In the Beginning 15

CHAPTER 2   The True Vine 35

CHAPTER 3   States of Nature 59

PART II **Ritual** 79

CHAPTER 4   Hey Bear! 85

CHAPTER 5   Let Food Be Thy Medicine 107

CHAPTER 6   Deepak Chopra's Condo 131

PART III **Law** 147

CHAPTER 7   The Invisible Hand 153

CHAPTER 8   The Rhythm 171

CHAPTER 9   God-Given Talent 190

Afterword: Salvation 206
Acknowledgments 214
Notes 216
Select Bibliography 235
Index 241

Introduction

HOW CAN WE LIVE IN HARMONY with nature?

Animal species, millions of years in the making, gone in a geological blink; ancient waterways clouded with waste; the land disemboweled and burned; the fumes heating the planet—with every passing day the damage grows worse, the question more urgent.

One promising answer was served for dinner at a recent conference on the future of environmentalism. Along with some of the world's foremost scientists, journalists, and academics, I bit into my Impossible Burger, the latest in high-tech fake meat, and wondered if it would live up to the hype. Not just in terms of flavor—which to my palate was indistinguishable from a standard fast-food patty—but as a solution for our broken relationship with the natural world.

Enthusiasts see meat substitutes as essential, even salvific. "Fake Meat Will Save Us," proclaimed a headline in the *New York Times* the morning after our dinner.[1] Impossible Foods' mission is "To save meat. And earth." Their enthusiasm is understandable: scientists agree that animal agriculture as it is currently practiced plays an outsize role in climate change, oceanic dead zones, antibiotic-resistant bacteria, and other unnatural horrors. Then there's the ethics of eating creatures bred to be cheap factory outputs, crammed cheek by jowl in ghastly living quarters until they're old enough to be slaughtered. Impossible Burgers and other fakes provide an appealing alternative, allowing us both to reduce our culinary carbon footprint and to treat animals with dignity, instead of seeing them, and the rest of nature, as raw material meant for the fulfillment of human desires.

Only days earlier, however, my family and I had dined on a radically different answer—a juicy pork loin purchased from Polyface Farm, just an

1

hour from our home. The farm is 550 bucolic acres of open grassland and wooded hillsides deep in Virginia's Shenandoah Valley, at the end of a bumpy dirt road called Pure Meadows Lane. The road was named by Joel Salatin, founder of Polyface, to reflect the natural purity of the land he stewards, and his redemptive vision of a purified relationship between humans and what he reverently calls "Creation."

Touring the farm I found myself wholly won over by the power of his vision, which the journalist Michael Pollan described as a "dance on the theme of symbiosis" performed by humans, animals, and plants on nature's stage.[2] Upon arriving I saw a toddler playing happily with rabbits in front of open greenhouses—not pet rabbits, but rabbits that would eventually be slaughtered and eaten or sold. With a few other visitors I hiked up a hill, stepped over a single electric wire, and joined a group of pigs that were rooting away with pleasure. We shared a field with cattle as they flicked flies with their tails and munched on native grasses. "Good farming should be aesthetically, aromatically, and sensually romantic," said Salatin proudly, surveying the scene. His farm certainly was. Birdsong and the hum of insects filled the sweet-smelling air, and I felt a primal need for closeness with nature fulfilled, as if I were on a hike in a national park.

A passionate environmentalist, Salatin regularly calls out the hubris of fellow conservative Christians who seek to replace a divinely ordained natural order with one of their own making. Unlike "chemical-wielding farmers," as he refers to them, Salatin raises chickens, pigs, rabbits, turkeys, sheep, and cattle according to "nature's template." Lagoons of manure that contribute to dead zones? Every ounce of manure at Polyface fertilizes vegetables and grasses, in addition to attracting grubs that feed the chickens. Routine dosing with antibiotics? Unnecessary, declares Salatin, because he allows his animals to live as nature intended.

According to his philosophy, Impossible Burgers epitomize the problem with our relationship to nature, not the solution. They depend on genetically modified soy for a protein called "soy leghemogloblin" (the eleventh of twenty-one ingredients), which lends a characteristic meaty flavor. "GMOs . . . egregiously violate God's pattern," claims Salatin, a phrase he uses interchangeably with "nature's pattern."[3] He has blamed the consumption of genetically modified foods for everything from increasing allergies to increasing violence. When I asked if he could imagine some role for these technologies in the future of food, his answer was unequivocal: "There is no redeeming value to GMOs or fake meat." If we want to live in harmony with Creation, we should honor the wisdom of natural systems rather than manipulate them further. Like ancient hunter-gatherers and traditional small-scale farmers, we must sanctify our relationship with the animals we eat, not remove ourselves from the cycle of life. The way of artificial hubris ends in damnation, not salvation.

Over our dinner of genetically modified patties I described Salatin's perspective to other attendees. Since the conference was sponsored by an organization that strongly supports technological approaches to conservation, their responses were unsurprising. Some scoffed at the very idea of a division between natural and unnatural. *Isn't everything natural? Isn't nature a social construct?* Others were willing to grant the division, but insisted that saving nature depends on transcending natural patterns. Denser cities, effective birth control, and intensive agriculture may violate "nature's template," but they will help us meet important conservation goals such as reduced land use and lower carbon emissions. The contrast with Salatin couldn't have been starker: when Pollan pressed him on how to feed New York City using locally sourced food, Salatin replied, "Why do we have to have a New York City? What good is it?"[4]

Pork loin or Impossible Burger? The country or the city? Again and again, the question of our proper relationship with nature is reframed in the same way: Should we obey nature or transcend it?

This book is a comprehensive response to that question. Instead of choosing sides, it shows how the framing is fundamentally misguided and counterproductive. An oppositional binary between "natural" and "unnatural" inhibits constructive dialogue about humanity's most pressing problems. It trades complicated truths for the comfort of clear categories. It encourages dogmatism over compromise, certainty over humility, and simplicity over nuance.

Breaking down the binary won't be easy, however, since the appeal to natural goodness is among the most influential arguments in the history of human thought. Far from being restricted to modern debates over food, we will see how this argument has influenced virtually every aspect of our individual and collective lives, from sexual habits to economic principles, from how we raise our children to how we organize their sports.

Variations on the idea of "natural goodness" are ubiquitous in all intellectual traditions, ancient and modern, East and West. In ancient China, sages wrote approvingly of *ziran*, the "self-so" of natural impulses uncorrupted by human meddling. Monks promised miraculous healing if people abandoned their homes for caves and gave up farming to forage in the wild. Across the globe, Greek poets told of a golden age when natural laws made politics unnecessary and a golden race lived effortlessly off nature's bounty. *Natural*: it describes Eden before agriculture, our instincts before we sinned. It is ideal beauty and the best sort of food; how we ought to act and how we hope to die—naturally.

Likewise, "unnatural" has always connoted evil. Throughout Shakespeare's plays, for instance, the word serves as shorthand for moral deficiency: "unnatural and unkind," "unworthy and unnatural," "savage and unnatural," "inhuman and unnatural." In his time—and long before it—an "unnatural" birth meant a baby born with some deformity; an "unnatural" death meant, and still means, life cut short by murder or sickness. With regard to sexual activity, "unnatural" describes perversions of desire; in government, perversions of justice.

Yet there are obvious contradictions built into this binary. Disease, deformity, and infant death are "unnatural" occurrences that feature prominently in the natural world, the very drivers of evolution. Violence and resource scarcity mark the natural existence of all animals, humans included. Those Chinese monks wrote their praises of natural living on unnaturally processed bamboo strips, just as advocates of natural living today spread their message via social media.

So how have people overlooked these contradictions and arrived at a vision of natural goodness and unnatural evil? Why is it that even some tech utopians insist on justifying technology by arguing that it, too, is natural, instead of simply embracing unnaturalness?

As a scholar of religion, I find the answer is clear. Another primary goal of this book is to show that despite appearances, *"natural" is a religious term*. The impulse that guides so many shopping carts and parenting decisions is thoroughly theological. "Nature" is another term for God; "natural," a synonym for holy. "Unnatural" acts are violations of nature's wise commandments, laws inscribed in the structure of reality and ignored at our peril. Only humans are capable of such violations, since by definition anything unnatural—synthetic, artificial, fabricated, manipulated, manufactured—is caused by us, not nature.

This way of thinking is extremely compelling. Even Charles Darwin was drawn to articulate his theory of natural selection as if it were orchestrated by a benevolent deity. Confronted by the painful realities of the natural world, Darwin admitted it was derogatory to suggest that "the Creator of countless systems of worlds should have created each of the myriads of creeping parasites and worms."[5] But in the very same passage he maintained that the laws of evolution "should exalt our notion of the power of the omniscient Creator." At other times he divinized nature directly, capitalizing Nature and speaking of her as a hidden goddess "silently and insensibly working . . . at the improvement of each organic being," slowly making "progress towards perfection."[6]

I began with the examples of pork and Impossible Burgers because I first noticed this unique form of religiosity while researching attitudes toward food. In 2016, the US Food and Drug Administration put out a call for public comment on the meaning of "natural" to help inform regulatory practices. It's an important question: data show that over half of Americans prefer "all natural" food. Market analysts have even found that "natural" is a more desirable attribute than "organic." Reading the 4,863 comments from the respondents confirmed to me that this preference was deeply religious. There were many references to nature's "intentions," as if a higher power *wanted* us to eat certain foods but not others, and had our best interests at heart. Some comments were straightforwardly theological: "Natural should be limited to those ingredients that have been created by God," said one.[7]

But despite its popularity, the idea of "natural food" is impossibly vague and relative. While people have long advocated for the benefits of eating naturally, there's little consensus on what it means. After all, the freakishly large vegetables and fruits at farmers' markets would be unrecognizable to early hunter-gatherers, and so would the carefully bred chickens and cattle raised at Polyface Farm. I came to believe that a clear-eyed understanding of our food system required abandoning historically naïve and theologically inflected visions of natural goodness.

However, my growing skepticism about natural goodness was accompanied by guilt and doubt. In a world wracked by global warming, with plastic bags and discarded tires polluting the seas; facing, as we do, an epidemic of extinctions and endless man-made threats to public health, wasn't it reckless, almost sacrilegious, to question the intrinsic goodness of nature and reverence for natural living? Our technological hubris seems uniquely responsible for the sorry state of the earth. Salatin belongs to a long tradition of environmentalists whose passion was fueled by belief in the divine wisdom

of natural systems. We owe them an enormous debt of gratitude for forcing us to recognize the importance—indeed, the sacredness—of the natural
world, and how humans were imperiling it. If skeptics like me had won the
day, their voices might have gone unheeded. Why push back on beliefs that
humble us and urge respect for the environment?

One reason jumped out immediately: unexamined faith in natural goodness can lead to tragedy. Reverence for natural approaches to health and healing caused the preventable deaths of hundreds of thousands of South Africans
from AIDS between 1999 and 2008.[8] They were advised by their president,
Thabo Mbeki, and his health minister, Manto Tshabalala-Msimang, to avoid
"toxic" chemical drugs and instead treat the disease naturally with beetroot,
olive oil, garlic, lemons, and African potatoes.[9] The ongoing public health
crisis of vaccination refusal has a similar basis in the ideology of natural purity. Parents who refuse vaccines tend to see them as unnatural interventions.
As with proponents of natural food, there is an underlying religious component to the parents' beliefs, occasionally made explicit through the conflation
of God and nature. "We have a God-given immune system," said one mother
in an interview conducted by the sociologist Jennifer Reich. "You don't have
to shoot things into the body. Let's support what we already have."[10]

Catastrophic medical decisions are only the beginning when it comes to the
distorting force of appeals to nature's intentions. Sexists invoked the natural inferiority of women to justify their exclusion from education, athletics,
and politics. Eugenicists used natural selection to rationalize their breeding
programs. Opponents of interracial marriage argued that unions between
different races were unnatural, and opponents of same-sex marriage make
a parallel case today. Ethnic nationalists secure followers by painting a nostalgic picture of once-great nations peopled only by their original, "natural"
occupants. (We see the residue of this in the legal language of "natural-
izing" citizens.) Advocates of radical economic deregulation have justified
their proposed policies by appealing to economist Adam Smith's naturally
emerging "invisible hand" that reaches down from the proverbial heavens
to guide ideal commerce—or would, if only unnatural political interference
were removed from the picture. On the other side of the ideological spectrum, communist revolutionaries argued that political violence was part of
the natural progression of history from capitalism to communism.

Although these arguments usually depend on flawed science, good science is no antidote. Justifying same-sex marriage by pointing to homosexual activity in primates, or to genetic determinants of sexual orientation, only repeats the fundamental mistake of seeking morality in the mirror of nature. What if homosexuality only occurred in humans? What if there were no genetic markers for sexual orientation? Could we then condemn homosexuality as unnatural? What if a nation was once populated by only one ethnicity? Should it therefore expel more recent immigrants? Natural goodness sounds great in theory, but in practice it's a mercenary ethic that anyone can hire to fight for their cause.

And then there's the problem of my own privilege. I have never walked for miles to secure drinking water, or medicine, or books. Machines wash my clothing and my dishes. Natural light is a wonderful thing, but for nearly a billion people with no access to power, and billions more with only intermittent access, the prospect of unnatural light is no less wonderful.[11] No one has told me that my sexual orientation is unnatural, or my biology is naturally inferior. Seen differently, praise of naturalness can look like praise of the privileges that render its downsides invisible.

This point was made eloquently to me by the late Calestous Juma, a professor of international development at Harvard University. As a young boy in rural Kenya, Juma's evening entertainment was often stargazing. Ninety miles from the nearest city, he looked up into a pitch-black sky, pupils unshrunk by screens or filaments or the glow of his family's kerosene lamp.

But there were also rare nights when the late arrival of a bus disturbed the stillness. First there was the distant growl of an engine, then headlights cutting through the darkness like a scythe through a grain field. His pupils contracted, the stars dimmed, and the natural world yielded to technology.

This moment can easily be seen as tragic: a scene of invasion, pollution, and desecration. In his classic paean to nature, *Walden*, Henry David Thoreau calls the local locomotive a "bloated pest," its iron belly full of mercenaries bent on ravaging his idyllic woodland.[12] And yet, any Kenyan child can tell you that Thoreau's perspective is not the only one. Rather than mourning the arrival of the bus, Juma described to me how he and other villagers would rush to the marketplace and greet it. The approaching headlights promised all manner of unnatural wonders: bicycles, books, radios, medicine. To him, these treasures represented the very realization of human nature, an inborn drive to discover and create and improve our lot in life.

But nuance is powerless against appeals to Nature (with a capital N). Like citing a holy text, they have incredible persuasive force, making it difficult to assess evidence objectively, and easy to oversimplify complicated issues like economics, immigration, and love. Yes, what someone calls "natural" may be good, but the association is by no means necessary, or even likely. The philosopher John Stuart Mill said it bluntly: "Nature, natural, and the group of words derived from them . . . are one of the most copious sources of false taste, false philosophy, false morality, and even bad laws."[13] (The "appeal to nature" is so frequently abused that it has earned the status of an informal fallacy.)[14]

It's impossible to capture the scope of the problem by looking at a single time period or a specific feature of human culture, which is why the following chapters are interdisciplinary and transhistorical. A broader perspective reveals that wherever the concepts occur—whether avowedly religious or supposedly secular—"natural" and "nature" are the idiosyncratic products of culturally specific ideologies, deployed to justify a wide range of inconsistent positions. Nevertheless, people persist in believing that their version of naturalness is an eternal truth, holy scripture dictated by Mother Nature.

As with all religious faiths, the consequences of belief differ dramatically from one person to the next. For the orthodox, nature worship is a persistent and conscious factor in every aspect of their lives. They constantly ask themselves, "What would nature want?"—in the voting booth and during the school selection process, while shopping for cookware and the food they will prepare in it. Like monks, some make enormous sacrifices, living completely "off the grid" or giving up clothing as part of the "naturist" movement.

For the majority of us, however, faith in nature's goodness only affects certain aspects of our lives—some food choices here and there, or maybe that expensive organic mattress for the baby's crib. Even so, it is important to recognize that the effects of a lukewarm faith are not limited to individuals' decisions. Nature worship infuses culture in myriad ways, its power hidden under the sediment of time and built into the unconscious demands of custom and language, making it all the more difficult to recognize and challenge.

Rejecting theological versions of nature does not entail embracing their opposite. We should not ignore the genuine wisdom of those who see divinity reflected in nature, since they have often learned a great deal from careful attention to natural systems. Appeals to natural goodness are not always fallacious, and categorically dismissing them is its own equally perni-cious form of faith. Sometimes unnatural solutions will be catastrophically misguided, sometimes not—but only by identifying and abandoning faith in nature's goodness, wherever it occurs, can we know when that's the case. Since this faith may reflect an inability or unwillingness to think outside one's privileged context, bracketing it is especially important when it comes to social and political policy. The best future for humanity and nature must be built on dialogue and evidence, not taboos and zealotry.

I hope this book can help us achieve that future. Unlearning the ortho-doxies of nature worship will be liberating—and not just from the guilt of feeding our children the occasional non-organic snack. It will allow us to seek complicated truths instead of being tied to mythic binaries. It will help us clarify why nature is sacred and worthy of protection. It will facilitate the identification of propagandists, bigots, demagogues, and marketers who wrap their rhetoric in the mantle of "what's natural." And it is necessary if we are to be good stewards of the earth's welfare, and good neighbors to each other. As Calestous Juma said to me: "Mythology gives us an excuse to abdicate our responsibility to the world. Instead, we must be courageous enough to play our role responsibly."

PART I

Myth

"One doesn't expect Dr. Frankenstein to show up in a wool sweater," wrote the political commentator Charles Krauthammer, ominously, in the March 1997 issue of *Time* magazine.[1] He was referring to British scientist Dr. Ian Wilmut, who eight months earlier had successfully created Dolly, the world's most famous sheep, by cloning her from another adult sheep's cell.

Krauthammer's criticism was unsparing. "This was not supposed to happen," he insisted. Dolly was "a cataclysmic" creature. But PPL Therapeutics, the company responsible for funding the science behind Dolly, was undeterred, and four years later produced five cloned female pigs.

Again, the news provoked outrage. Lisa Lange, a spokeswoman for People for the Ethical Treatment of Animals, echoed Krauthammer when she dismissed justifications of cloning: "There's always a reason given to validate these Frankenstein-like experiments."[2]

Invoking Mary Shelley's *Frankenstein* is standard fare in arguments over controversial science. Five years before the creation of Dolly, Boston College English professor Paul Lewis coined the term "Frankenfood" in a letter to the *New York Times* that argued for stricter FDA regulation of genetically modified foods. "If they want to sell us Frankenfood," he wrote, "perhaps it's time to gather the villagers, light some torches and head to the castle."[3] And in 1978, in vitro fertilization pioneer Dr. Patrick Steptoe tried to preempt such criticism when he defended his role in the birth of Louise Brown, the world's first "test-tube" baby. "I am not a wizard or a Frankenstein," he pleaded.[4]

Steptoe was wise to dissociate himself from Frankenstein. The story is particularly potent, deftly incorporating elements of the world's most famous myths. The subtitle of the book is "The Modern Prometheus," a reference to the ancient Greek Titan who stole fire from the gods and ended up chained to a rock, his liver to be devoured daily by a vulture and regenerated at night in eternal punishment. And who can fail to notice the resemblance to the Fall: temptation leads Adam and Eve, like Dr. Frankenstein, to acquire forbidden knowledge, which results in their expulsion from Eden. Their punishment included the curse of having to work the land—agriculture—and, for Eve, pain during childbirth and enforced patriarchy ("he will rule over you," Genesis 3:16).

People don't just live by archetypal myths—they are constituted by them. As the scholar of myth Bruce Lincoln has argued, myth is *ideology in narrative form*. Want to know what someone stands for? Look to the myths that shape them. Group identity, from religion to politics, depends on an investment in a few foundational stories, which serve as justifications of one's preferred moral and social order. This is why mythically justified beliefs are so resistant to evidence: changing them means changing yourself.

Another way to think about myths is as narrative metaphors. Metaphors are indispensable for our understanding of reality, and shared metaphors are threaded through the fabric of a culture. In their seminal work on metaphor, the linguists George Lakoff and Mark Johnson show how metaphors like "argument is war" shape everyday language, and, in turn, our perspective on what is metaphorized: He *attacked* my arguments. She *shot down* my main points. Your argument is *indefensible*.[5] When Krauthammer says that Dr. Wilmut is Frankenstein, he activates a set of implicit claims about cloned animals, all derived from the story: Dolly is a *monster*. She is *out of control*. She is a *danger*.

Calling something natural may seem like a straightforward description, but it is actually a powerful narrative metaphor. The adjective "natural" cannot be understood without reference to "nature," a concept that necessarily depends on a narrative. Nature is not a static object, but rather a story that takes place over time. The *Oxford English Dictionary* defines "natural" as "something having its basis in the natural world or in the usual course of nature." But what exactly is "the usual course of nature"? When did it begin, how did it proceed, and where do we locate departures from it? Answers to these questions come in the form of myths, the stories that we depend on, consciously or unconsciously, whenever we judge a product, a process, or a way of life to be natural.

In a sense, then, "natural" is better understood as a meta-myth. It is built from multiple archetypal narratives that come together to form an overarching conceptual framework. We can use this framework to evaluate anything. Food judged to be natural is food that fits certain story lines, ones we see reflected in packaging design: red farmhouse, green leaves, animals grazing, sun shining, fresh fruits and vegetables, a grandmother with a wooden spoon. These story lines are deeply ideological. Just consider the common pairing of "natural" with "healthy," "true," "pure," "honest," "authentic," "simple" and "real." At the sight of the word "natural" we reflexively tell ourselves myths about *authentic* culinary traditions handed down for generations, or *honest* farmers who work in sun-dappled fields to bring us whatever *real* grocery store product has chosen to incorporate the power of these narrative metaphors on its label.

Labeling a story as "myth" might seem like a pejorative judgment of falsehood, but myths are not necessarily bad. Just as "mythic" can mean either "false" or "awesome," myths themselves may be harmful superstitions or expressions of sacred, primal truths. Nor is the veracity of a myth, strictly speaking, the most important thing about it. Like *Frankenstein*, George Orwell's *Nineteen Eighty-Four* achieved mythic status, to the point that "Orwellian" has become an adjective commonly used to describe actual political policies. While based on Orwell's lived experiences, the book is, of course, still a fiction, but one taken to provide wise guidance about the real-world dangers of totalitarianism. Truth is important, but figuring out which myths to believe in—that is, which myths should shape your ideology—is more than an exercise in fact-checking.

Over two thousand years ago Plato recognized this ambiguity, writing that myths "on the whole are false, but still they have some truth in them." Famously suspicious of poets and their fictions, he nevertheless saw these narrative metaphors as essential to the education of good citizens. "One must supervise the makers of myths," he argues in *The Republic*, "and one must approve that which is good in their compositions and condemn that which is not. And those which are approved, we will persuade nurses and mothers to recount them to their children, and to shape their souls with these myths, even more powerfully than they shape their bodies with their hands. But one must reject most of the stories they now recount."⁶ Since we are not subject to the whims of a philosopher-king, we must be our own supervisors. We must decide which myths to tell ourselves, our friends, and our children, a process that starts with becoming conscious of the myths that influence our individual and collective lives.

Our understanding of what's natural can never be completely free of myth. Nature is inherently mythic, a concept so gigantic in scope and significance that only with narrative metaphors can we conceive of it, speak of it, at all. But myths can be interrogated, measured against reality, and scrutinized for alternative interpretations and implications. We can trust Krauthammer on the meaning of *Frankenstein*—it's a horror story about humans disturbing the natural order—or we can judge it for ourselves.

In the original tale, Dr. Frankenstein's creation is no monster, but rather a kind, gentle Creature. Tragically, the Creature soon learns to fear the humans he loves, who, terrified by his appearance, reject him and never come to understand his true identity. The real villains in Shelley's story are neither Dr. Frankenstein nor his artificial creation—or at least not *just* them but also the intolerant, torch-wielding villagers. Only after experiencing cruelty does the Creature become a monster, exacting revenge on those who refused to give him a chance. Read in this way, the novel suggests a radically different vision of our ideal relationship to the natural order than the more familiar version.

The next three chapters critically examine some of the most important myths we tell ourselves about nature, from the epic to the everyday. At times this examination will feel profoundly disorienting—at least, I know it has felt that way to me. Changing our understanding of myths means interrogating ourselves, and no myths are more important to us than those about nature. But we don't have the luxury of sticking to the status quo. To invoke an apocalyptic myth that has some truth in it: Judgment day is never too far off, and the future of the natural world depends on our ability to examine myths wisely, to approve that which is good and condemn that which is not.

CHAPTER I

In the Beginning

IN THE BEGINNING there are birth stories.

They appear in virtually all known cultures, always with mythic significance. Where did we come from? Hopi Native Americans remind themselves with *sipapu*, symbolic holes in the floor of ritual chambers that recall our emergence from earth's warm womb. Why are we here? Astrologers read heavenly meaning into human lives by linking the circumstances of one's birth to the stars, as did Shakespeare: "From forth the fatal loins of these two foes / A pair of star-cross'd lovers take their life." How to explain tragedy? Blues musicians answer with the same cosmic birth myth that echoed in the temples of ancient civilizations: "Born under a bad sign." How to escape tragedy? The Gospel of John: "No one can see the kingdom of God unless they are born again."

Humanity is fascinated by origin stories because they seem to disclose essential and prophetic truths. They tell us, individually and collectively, what we really are and where we are ultimately headed. Like the Ouroboros serpent eating its tail, the end is hidden in the beginning. The savior is birthed by a virgin. The newborn phoenix is dusted with ashes. Birth stories—creation myths—are metaphors with a plot, manageable visions of massive timescales, overwhelming contexts, the vast and mysterious systems that inexplicably begot life. *How did I get here?* Mother Earth, Mother Nature, the Milky Way that burst from Hera's breast. The Hebrew word for human, *adam*, comes from "earth." "By the sweat of your brow you will eat your food until you return to the ground, since from it you were taken; for dust you are and to dust you will return" (Genesis 3:19).

What we are made of is where we will go, which is whence we came.

The existential weight of these narratives explains the allure of genealogy. It is as if some secret essence is slowly revealed by moving backward, birth after birth, ancestor after ancestor, until genealogy gives way to anthropology and we are seeking ourselves in the mirror of prehistoric humans, back further, still, to evolutionary biology and the animals from which we are descended, back, finally, to whatever forces led to life instead of matter, order instead of chaos, something instead of nothing. We are defined by our genes, our birthplace, our generation, the elements of our genesis.

A similar revelatory power belongs to the origin stories of words, their etymologies. The etymology of "etymology" is knowledge (*logos*) of truth (*étumon*). To know the origin of a word, like knowing the origin of the universe or a superhero, is to know an essential aspect of its meaning.

The word "nature" has its genesis in the Latin verb *nāscor*, "to be born," a linguistic seed that grew into "natural," "naturally," "nation," and "native," with all their associated meanings of truth, essence, authenticity, and belonging.

In an etymological sense, there is nothing more natural than birth, and in a mythic sense there is nothing more significant.

My mom wanted a natural birth, but she did not get one. I ended up four weeks late in frank breech presentation, feet by my ears and arms crossed as if refusing to enter the world. Her doctor recommended a caesarean, standard in such risky situations.

"I was disappointed," she told me, "but also grateful to be living in modern times, to be safe, and for you to be safe." She paused for a moment. "And I got to have a natural C-section, where I was awake and saw you come out."

Even in 1981, the ideal of naturalness, with all its mythic majesty and paradox ("natural C-section"?), featured regularly in descriptions of childbirth. In the 1950s, Dr. Fernand Lamaze captivated the world with his admiring firsthand accounts of midwifery as he saw it practiced in the Soviet Union. A few decades later it was everywhere. My mom and dad took Lamaze classes where a central tenet was, and remains, the superiority of natural birth. From the official website: "Lamaze International is a nonprofit organization that promotes a natural, healthy and safe approach to pregnancy, childbirth and early parenting."[1] Not only is *natural* a core part of their identity, but here it takes phrasal pride of place, before *healthy* and *safe*.

As any parent knows, the value of being natural is omnipresent in the world of birth advice. Best-selling books inveigh against a mechanized medical establishment that eschews natural (also known as "normal") birth for Pitocin-induced labors, unnecessary caesareans, and analgesic cocktails that commonly include fentanyl, the newly infamous villain of the opioid epidemic. When we were soon-to-be parents, my wife and I felt suddenly compelled by the idea of natural goodness, and I've found other parents report the same feeling. The impending birth of a child confronted us with a facet of being human that usually remains in the background. We became profoundly biological: my hand on her swelling belly, she *with a baby growing inside her*, the baby's kicks testifying to an unbelievable process that was unfolding, without conscious intervention, as it has for hundreds of thousands of years.

My wife never once considered going without Novocain for a filling, but she did consider laboring without painkillers. I didn't want her to suffer, but I understood her desire. What drove it, exactly? Neither of us knows for certain, but it came from somewhere deep and pre-rational, where you make meaning out of life's mysteries. We wanted our child's origin story to be perfect and pure, my wife an avatar of the creative force responsible for life and humanity. For us, like so many others, that meant wanting birth to be natural.

My mom also remembers objecting to an "unnatural" epidural. Maybe it's because her Lamaze coach told her some variation on the following, also from the official website: "The pain of labor and birth, like other pain, protects us."[2] Had she been pursuing the Bradley Method of Natural Childbirth, another popular option at the time, she might have thought that natural labor—that is, labor without medication or medical intervention—is usually painless. Does the laboring mother have an agonized expression on her face, despite perfect adherence to Bradley's recommendations for diet, exercise, and deep breathing? "Effort, not pain," Bradley reassures his readers. "Please, please do not confuse the two."[3]

Bradley's book—still selling briskly after over a half-century—weaves biology and theology together into a divinized narrative of natural childbirth. He takes pains to clarify that the Bradley Method did not originate with him, but with a higher power, whose instructions are written in the Book of

Nature. "It is truly God's method," writes Bradley, "the way He taught it to animals via inner instinctive drives, and the way we human beings must learn to do it by imitation."[4]

In "God's great outdoors," animals give birth without hospitals or doctors, without fluorescent lights and latex gloves, and always (in Bradley's account) without grief or difficulty. We modern humans have fallen from that world, tempted by drugs and technology, and the result is sickly children and unhappy adults. Natural childbirth offers the redemptive possibility of "beautiful, healthy, Bradley babies" who will "solve the problems of the world with brains unaffected by drugs at birth," born according to eternal principles "that have been tested from ancient times to the present."[5]

When God drops out of natural childbirth books, He is most often replaced by nature, occasionally veiled in the passive voice. "Mother Nature designed it this way" (*The Mama Natural Week-by-Week Guide to Pregnancy and Childbirth*); "beautifully and admirably designed to give birth" (*Ina May's Guide to Childbirth*); "nature has a perfect plan" (*The Official Lamaze Guide*).[6]

One evolutionary biologist I spoke with insisted on correcting "we were designed" to "we evolved," because she was concerned that design implies intention—what is natural is *good* because the (implied) Designer made it that way. Hers is a losing battle. The linguistic conceit is ineradicable from colloquial speech. Even books that don't favor natural birth still talk about nature as an agent with our best interests at heart, as in these passages from *What to Expect When You're Expecting*, the most successful childbirth guide ever written: "nature's incredibly good at what it does"; "nature knows a thing or two about nutrition"; "stick with nature's model when you can"; "nature's way of preparing you"; "nature has your baby's back."[7]

If nature has your baby's back, it follows that women's inborn biological wisdom should serve as the primary guide to childbirth—which, these guides tell us, contrary to what obstetricians would have you believe, is not especially dangerous. A common argument among natural birth advocates is that having babies in the distant past was actually less dangerous than it is for women living in modern industrialized society. "With the birth process, it's how nature intended it to be," said Debbie Wong, a Charlottesville, Virginia, midwife who had helped one of my colleagues with her birth. Inborn wisdom was a prominent theme of our conversation: "I believe in every woman's ability to give birth, just as it has happened for millions of years," she said. "I don't believe women have stopped being able to give birth."

For Wong, sickness, death, and unendurable pain are only rarely part of the natural birth process, and far more often the result of rejecting it. *Trust your body* is a frequent refrain in guides to natural birth. Medicalization subverts this trust, turning the miracle of birth into a disease and recasting the life-creating mother as a sick patient, all to the advantage of a technocratic establishment.

Natural birth advocates see themselves as offering an alternative to medicalized birth stories. Birth, as they tell it, is not the enemy, to be monitored and fought as if it were a tumor, when really it's a child growing in the mother's womb. What we should be fighting is the transformation of childbirth into a dehumanizing experience, infused with fear and drained of symbolic power. Natural birth promises a better mythic narrative. Mothers are at once the inheritors of nature's greatest power, and, in the experience of actively bringing new life into the world, mother goddesses themselves. "In this story of birth, labor unfolds quite naturally," reads the Lamaze guide. "No machines time or rate your progress. No experts tell you what to do or what not to do. No one takes your baby from you. It's an ancient story of a strong confident woman and a competent baby who both know just what to do."[8]

The internet is filled with testimonials to this vision of birth. Perhaps most famous is a viral video with over seventy-nine million views, in which Simone Thurber, a thirty-nine-year-old Utah doula, gives birth at an isolated creek in the Australian Daintree rainforest. She had been inspired by her parents' stories of natural births that they saw as missionaries in Papua New Guinea. "I remember mum often telling us when a local lady had her baby and how at the time when she was due to give birth she simply found a nice spot, pushed out the baby, breastfed it and wrapped it in a carrier cloth, tied the child to her back and went about her business," recalled Thurber in an interview. "That stuck with me and is one of the reasons I wanted to give birth to Perouze [her daughter] in the wild."[9]

The popularity of Thurber's video inspired a Lifetime TV show, *Born in the Wild*, in which expecting mothers renounce the structures and strictures of modernity to embrace the great outdoors. "Modern parents, giving birth in the wilderness, like their ancestors," promises the male narrator at the beginning of episode 1, "Alaska: Remote and Unassisted." It is the promise of myth made real—a glimpse at how the natural goodness of authentic birth can be reclaimed.

A skeptic is likely to remark on the natural birth movement's paradoxical embrace of birth guides and midwives. The historical ubiquity of childbirth assistance means you can't really trust nature, right? To this, supporters of natural birth will reply that the role of traditional birth assistants is to facilitate nature, not improve on it or usurp its place. Whereas obstetricians *deliver* babies, midwives say they *attend* the pregnancy and *catch* the baby, foregrounding the mother's agency. The mother—and thus, by proxy, nature "herself"—is the central active participant. Midwives do not impose their own schedule with due dates, inductions, or unnecessary C-sections, instead serving a process that unfolds organically. On rare occasions interference may be called for, but only of the most minimally intrusive variety, with no need for scalpels, forceps, stirrups, or painkillers, and always with an attitude of reverence. In a revealing anecdote, the American midwife Ina May Gaskin recalls asking indigenous Guatemalan midwives how they learned a technique for dealing with shoulder dystocia, when the baby's head emerges but the shoulders remain caught: "The oldest of them pointed to the heavens and said, 'Dios. We learned it from God.'"[10]

Had she said "nature," the meaning would have been identical.

To further emphasize the intrinsic goodness of natural birth, advocates like Gaskin blame typical birth complications on unnatural living. Daily consumption of processed foods, countless hours spent hunched over a keyboard with hips squeezed into an office chair—these are identified as violations of nature's laws, for which women are sentenced to more difficult births. The logic has a beguiling simplicity, particularly if you've benefited from a standing desk or had surgery for carpal tunnel syndrome. "If you reflect on Nature's plan for a moment you will see that she doesn't want any of her children to suffer from any type of physical disease," writes Ramiel Nagel in *Healing Our Children: Sacred Wisdom for Preconception, Pregnancy, Birth and Parenting*. "If Nature had designed humans or animals to suffer from disease, then we couldn't have survived as a species so long." The inexorable conclusion is that pregnancy complications, like all health problems, are "the tragic results of our modern lifestyle."[11]

Historical accounts of birth in non-European cultures are often cited as corroborating evidence of our collective fall from grace. In a classic medical treatise, the physician and founding father Benjamin Rush described Native American women's birth practices in the same terms that Simone Thurber's mom described the natives of Papua New Guinea. "Nature is their only midwife," observed Rush. "Their labours are short, and accompanied with little pain. . . . After washing herself in cold water she returns in a few days to her usual employments."[12] The eighteenth-century French polymath Comte de Buffon was equally admiring of indigenous African women: "[They] bring forth their children with great ease, and require no assistance. Their labours are followed by no troublesome consequences; for their strength is fully restored by a day, or at most two days repose."[13] In 1865, just six years after Darwin published *On the Origin of Species*, the biologist Thomas Huxley wrote that "the bearing of children may, and ought, to become as free from danger and long debility to the civilised woman as it is to the savage."[14] (Huxley is quoted approvingly in *Ina May's Guide to Childbirth*.)

In retrospect, however, much of this looks like ideologically driven fairy tales. In the time of Benjamin Rush and Comte de Buffon, tales of "painless parturition" were used to dehumanize indigenous peoples and to justify working enslaved women as hard as men, even during pregnancy.[15] They became animals that had escaped Eve's curse, which made it easier to see them as biological oddities meant for freak shows. "Hottentot Venus!" read advertisements for the popular traveling exhibition of Sara "Saartjie" Baartman, a Khoikoi woman of southwestern Africa who, upon her death in 1815, was dissected by the French zoologist Georges Cuvier. He put Baartman's brain and genitalia on display in a museum as examples of an "ape-like" creature's biology. Post-Darwin, stories of easy birth worked to buttress theories of evolutionary superiority that ranked races on a spectrum of "less" to "more" evolved, the former to be analyzed by the latter in books with titles like *The Backward Peoples and Our Relations to Them* (Sir Henry Johnston, 1920).

Animalistic, uncivilized, resistant to pain. Today, such distinctions can be seen for what they often were—convenient justifications of colonial atrocities and systemic racism. (The latter, in its current medical incarnation, results in black patients receiving less pain medication than white patients for the same conditions.)[16] Easy indigenous births are, on this reading, nothing more than harmful myths, wish-fulfillment in narrative form that has always served the powerful. As the political scientist Candace Johnson writes:

> The fantasy of Third World women's natural experiences of childbirth has become iconic among first world women, even if these experiences are more imagined than real. This creates multiple opportunities for exploitation, as

reports. "The opposite was true: the animal mother's eyes were radiant with joy and happiness."[18]

Bradley is not alone in his idealization of animals. Ina May Gaskin makes an implicit case for the ease with which animals give birth by explicitly invoking the divine wisdom that designed all life: "Remember this, for it is as true as true gets: *Your body is not a lemon.* You are not a machine. The Creator is not a careless mechanic. Human female bodies have the same potential to give birth as well as aardvarks, lions, rhinoceri, elephants, moose, and water buffalo."[19]

It really would be extraordinary if post-industrial humans were the only animals that struggled with birth, a strong argument for sticking with nature's design. But Bradley's and Gaskin's assertions are theological, not zoological. Conveniently, Gaskin does not mention the spotted hyena, a species "designed" so that females give birth (and urinate) through a penis-sized clitoris. When primiparous hyenas push their first pup down the birth canal, that clitoris splits open like a banana peel, resulting in high rates of infection and an estimated maternal mortality rate as high as 10 percent. (By way of comparison, Sierra Leone's human maternal mortality rate, the highest in the world, sits at "merely" 1.36 percent.)[20] The journalist Nathanael Johnson wryly captures the theological implications: "The fact that any animal must give birth this way seems the final proof that if nature had a designer, He—it seems unlikely, in this case, to have been a She—was vindictive or willfully negligent."[21]

Gaskin does mention lions, which have a neonatal mortality rate of 29 percent. *Over one in four lions dies before reaching one week old.*[22] In Afghanistan, the most dangerous country to be a baby, infant mortality is "only" around 11 percent, a statistic that includes deaths during the entire first year, not just the first week. Comparably horrific rates are reported in domestic animals—13.7 percent of horses are stillborn; 8 percent of puppies die within the first week—which are impossible to square with Bradley's report of "no objective evidence of pain or suffering." (That's not to mention the poor hyenas, whose neonatal mortality rate is a ghastly 60 percent, with many of the pups asphyxiating while stuck inside the peniform clitoris after placental detachment.)

Even primates, our closest animal relatives, fare poorly. Among captive primates, neonatal mortality is extremely high: 14 percent for gorillas, 20 percent for orangutans, and 25 percent for gibbons. I thought these high rates might be due to the "unnatural" conditions of captivity, but primatologists have ample evidence that birth complications happen in the wild, too. It's somewhat rare to observe feral primate births, since these usually occur under cover of darkness and in hiding, to avoid predation at a vulnerable time for mother and child. Yet over the years, painstaking fieldwork has led to significant findings. The first-ever observation of a feral baboon birth, for instance, happened in 1970, in Tanzania—and the infant was stillborn, in breech presentation, the same position I was in. During labor the baboon grimaced repeatedly, an expression usually associated with pain and fear. At last, the dead infant was delivered, whereupon the mother held the corpse to her chest while chewing on the placenta. The field notes are heartbreaking:

> RO [the mother's name] lays the infant on the ground in front of her, looks intently at it and grooms it.
>
> She lays the infant down, picks it up, sniffs it again.
>
> She appeared confused by the lack of movement and clinging which would have been present in a normal infant; she had born two live infants previously. She was later observed to place the infant on the ground and pull at its arms and legs repeatedly.[23]

There have been other reports of breech deliveries in feral primates, along with protracted labors and stillbirths. In 2017, researchers published details about the single largest data set of wild nonhuman primate births, observed in Ethiopian gelada monkeys over the course of ten years. Of fifteen births, two were stillbirths, a rate of 13 percent.[24]

Wild animals, domestic animals, non-primates, primates: the evidence is clear. Natural births are not free from complications. Asserting otherwise may be persuasive, but it is a fantasy that passes off Edenic myth as biological truth.

Turning to humans, there is little to indicate that "traditional" or "natural" birth is anything approaching what someone reading these words would consider safe. Among the Agta, a hunter-gatherer people in the Philippines, maternal mortality has been estimated at over 2 percent, worse than that of any country in the world.[25] For the Hiwi of South America it's slightly lower, at 1.3–1.8 percent, and for the Aché of Paraguay it's lower still, only 0.6 percent—which, though twice as good as Sierra Leone's, is still forty times higher than that in the United States, widely acknowledged to have the worst maternal mortality rate in the developed world.[26]

Like maternal mortality, infant mortality is variable for hunter-gatherers—since "they" are actually a diverse set of peoples who live in dramatically different environments with different lifeways—but it never drops below 15 to 20 percent. These high rates of death are reflected in social customs. The South Indian Nayaka and the Aka of the western Congo basin, for example, typically do not name babies for at least a year, and anthropologists have documented simple, unceremonious burials of newborns, as well as belief in the rebirth of babies as other babies.[27]

Experts I spoke with were unanimously opposed to portraying hunter-gatherer births as idyllic. "It's very common for women to have spontaneous abortions, and it's not unusual to have a stillbirth—to be born dead on arrival or to have the cord wrapped around the neck," said Barry Hewlett, an anthropologist who focuses on childhood and disease, and has worked extensively with the Aka. "From my own personal experience I know it's ridiculous to say there aren't problems."

Melvin Konner, a leading researcher of childhood in a cross-cultural context, agrees with Hewlett. Konner, both a physician and a biological anthropologist, is far from a cheerleader for modernity. He was among the first scholars to seriously consider the potential wisdom in hunter-gatherer childcare practices, as well as an early advocate of breastfeeding and eating more like our Paleolithic ancestors. But when it comes to modern obstetrics, he is unequivocal. "Obstetrics is one of the biggest success stories of medicine," he told me. "One of the great ironies is that it is precisely because of the success of obstetrics that you can have these movements like water births and homebirths and Lamaze and all the other cults. They wouldn't exist if OBs hadn't been so immensely successful that a lot of couples no longer think of pregnancy and childbirth as dangerous."

It's important to note that modern hunter-gatherers are not "living fossils." They struggle with displacement, the destruction of ancestral lands, and the introduction of previously unknown infectious diseases. (Studies of hunter-gatherers often qualify statistics or observations by saying they are "post-contact.") Yet archaeological evidence does not paint a substantially rosier picture of childbirth. Young adult females are disproportionately represented in the prehistoric fossil record, and researchers attribute part of that to birth complications. In a study of pre-Columbian female mummies from Arica, Chile (ranging in age from 1300 BCE to 1400 CE), 14 percent were found to have died from childbirth complications.[28] Another study of teeth that belonged to preagricultural people from Chile's Atacama Desert concluded that "in utero and postnatal stress were a common part of the infant experience in the Atacama."[29] And the earliest documented human twins—discovered in a giant Siberian hunter-gatherer gravesite dated to c. 8000 BCE—were not adults, but full-term fetuses, found nestled in the pelvic area of a young adult female. Apparently all three were victims of a birth gone wrong in times long past.[30]

None of this is surprising to those who study the evolutionary biology of childbirth. "I use childbirth all the time as an example of 'not-intelligent' design," said Karen Rosenberg, a biological anthropologist and leading expert on the evolution of childbirth at the University of Delaware. "It works, but just barely. There's nothing perfect about it. Like many other aspects of our biology, it's the best jerry-rigged thing that we've got." Birth is dangerous, difficult, and painful, by nature.

When nature is a benevolent God, these unpleasant truths about birth disappear into a universalizing faith. The perfection of natural design becomes non-negotiable, and we must blame ourselves (or our fallen culture) for any malfunctions. But nature is not perfect. The gears of evolution are greased with pain and death. As Annie Dillard writes in *Pilgrim at Tinker Creek*: "Evolution loves death more than it loves you or me. This is easy to write, easy to read, and hard to believe. The words are simple, the concept clear—but you don't believe it, do you? Nor do I. How could I, when we're both so lovable? Are my values then so diametrically opposed to those that nature preserves? This is the key point."[31]

To believe Dillard—and for me there is no other honest option—is to shatter the myths of perfect childbirth and painless parturition, of *natura* designed flawlessly by nature for our benefit. No matter: from the fragments, we can construct a new story about nature and natural birth, one that bears the marks of its genesis but does not share in the sins of its parents.

If this were a book about the science of "natural," now would be the time for a detailed evaluation of the science on C-sections, inductions, and epidurals, and a comparison with natural birth in terms of medical risks and benefits. But this is a book about the significance of "natural," not just the science of it, so first we will return to beginnings, the mythic power of origins and the meaning of natural birth.

Again and again, the women I spoke with who chose natural birth ex-pressed their preference in what might best be described as spiritual or po-etic language. For them, the value of natural birth couldn't be located solely, or even primarily, in scientific criteria such as a lower risk of vaginal tearing or the advantages of vaginal bacteria transfer to the newborn (though these were also among the benefits they cited).

"I was connected to women in a very different way," explained Hannah Smith, a sociologist at my university who chose a local midwife and certified OB/GYN to help her with unmedicated home delivery. "Having had this experience come through me, I feel connected to womankind, to my fore-mothers," she said. "And I felt breathtaking awe that I, my body, by itself, could do this incredibly dramatic thing."

Hannah is not religious, but being pregnant and becoming a mother had a kind of significance for her that seemed to call for religious terms. "There was a spiritual aspect to it, even though I don't believe in God," she told me, haltingly, as if words themselves were inadequate to describe the experience. "Here was my body, making another human being, and another human be-ing is more than just tissue and matter. And the process of bringing that be-ing into the world out of my body, there's something special and unknown and beyond the physical in that."

Her account resembled many of the birth stories that appear in natural birth guides, down to the theological overtones. "Someone once said that having a baby is our way of assisting God in a miracle," writes a woman named April in the Lamaze guide. "The moment [my son] was born, I real-ized what a miracle he was, what a miracle conception and birth are to us as parents and to humankind. I felt a kinship with every woman throughout history who has given birth."[32]

In the context of a hospital, Hannah feared this transcendent signifi-cance could fade into the background. A medicalized experience meant an experience that reduced pregnancy to a medical condition, described pri-marily as a risk, not a miracle, in an expert vocabulary not her own. There was no mystery, no awe, no veneration of the extraordinary process that was taking place inside her. No room for connection to primal forces or profound mysteries. Instead, the power, the agency, and eventually the baby himself would end up first and foremost in the hands of medical profession-als. It would be they, not she, who were ultimately responsible for bringing her son safely into the world. And that world? It belonged to them in every respect, from the physician's bland jargon to the sterile walls of the hospital where the birth story would be set.

Giving birth outside of the medical framework allowed Hannah to reclaim the miraculous dimension of the birth process—which, for her, was tied to having a natural birth experience. The passing of a child through her birth canal as other children had passed through other birth canals placed her in a lineage that stretched back into the mists of prehistory. In this sense, natural birth functions as a potent ritual that connects a specific mother with mothers as a natural kind, across cultures and back through endless generations. We were made by something beyond ourselves, but which is also a part of us.

We have all experienced the mythic significance of naturalness, even if we have not all given birth naturally. At the zoo, I see fragments of the natural world pacing unnatural enclosures. I am awed by what nature has birthed; I feel it is tragic that the animals are not in their natural environment. That tragic sense persists despite knowing that many animals live longer in captivity than in the wild. The loss cannot be reduced to mortality metrics—it's about the abrupt interruption of an ancient story. Likewise, when I take hikes it feels like they haven't truly begun until the parking lot disappears and the sound of automobiles has faded. The cars don't pose a risk to me. There is no way to quantify why I want them gone. But outside the language of quantification it's straightforward. They interfere with the meaning I seek, a private audience with nature.

Speaking frankly about the irreducible value of experiencing nature, whatever the context, does not make us irrational mystics—it makes us honest. When people protest the "medicalization" of birth, part of what they are protesting is the relentless push to translate the value of an experience into purely scientific metrics, a task as futile as it is depressing. Just ask Holly Dunsworth, a paleoanthropologist who specializes in the evolution of birth, and has proposed innovative new ways of understanding when and why birth became difficult for hominids. When she became pregnant, Dunsworth wasn't just thinking about risk reduction. A natural birth was a chance for her to experience what she studied, bodying forth what she had only theorized, the latest bud on the evolutionary tree to which she had dedicated her career.

"When I was a brand-new snarky grad student, I turned up my nose at the natural people," she told me. "Dream catcher, wacky, hippy-dippy bullshit. Risky." But then she got pregnant. And then she was told she was high-risk and would need a caesarean. In the doctor's office, there was no real room to articulate the importance of a natural birth. She had no choice, because in the cost-benefit analysis of birth, the currency of experience (beyond that of physical pain) has little value. "It was extremely frustrating and painful," recalled Dunsworth. "There was no way to communicate what this meant to me, and what I would lose if I didn't have it."

The tyranny of scientific metrics works in both directions. When a mother is told that natural breastfeeding is better for her child, it is hard to push back by citing the value of convenience, of being able to go back to work immediately, of not feeling tied at all times to your baby—in other words, the value of freedom, which, like nature, is another irreducible source of meaning, difficult to define but no less real for it, an important part of being fully human that does not line up neatly with what is most healthy. The essayist Eula Biss captures this tension of irreconcilable values in a poignant reflection on children and birth:

> I carry my son on the back of my bicycle and allow him to sleep in my bed, despite public service posters of a baby sleeping with a butcher knife that warn me, "Your baby sleeping with you can be just as dangerous." The disregard for statistical risk that researchers observe in people like me may be at least partly due to an unwillingness to live lives dictated by danger. We sleep with our babies because the benefits, as we see them, outweigh the risks. The birth of my son, which posed a greater risk to my health than I anticipated when I became pregnant, gave me a new appreciation for the idea that there are some risks worth taking.[33]

Like Biss, all parents are constantly balancing multiple values: the value of their own freedom, their children's freedom, of joy and safety, of novelty and security. For some mothers and families—for my wife and me—the potential for connecting with nature during birth is ultimately outweighed by other values, not least the reduction of pain and the sense of security that comes with a hospital birth. (This sense of security can be a function of class and race: healthcare systems do not distribute their benefits equally.) And that's fine. Not everyone connects to naturalness in the same way. Some people camp under the stars; others prefer the lodge. Some choose to live in the countryside surrounded by trees; others make their home in concrete jungles. Just because there is value in nature doesn't mean we should commune with it at all times, and castigate ourselves for the times we don't.

A middle way is also necessary in science, where nature plays a key role in the process of generating and testing hypotheses. In the evolutionary biology of childbirth, "nature" is a stand-in for what is sometimes referred to as the "environment of evolutionary adaptedness" (EEA). Originally coined in 1969 by the psychiatrist John Bowlby, the concept is fairly straightforward.[34] All species, including humans, are the products of evolutionary adaptation to a certain kind of environment. Remove a species from that environment and you might see biological dysfunction, or "evolutionary mismatch." In humans, evolutionary mismatch has been plausibly blamed for everything from rising obesity rates (we evolved in an environment where calories were not plentiful) to allergies (artificially hygienic environments are changing our microbiome and causing autoimmune disorders). However, it's unscientific to confuse the EEA with Eden, which would make any departure from one's original environment necessarily bad. This decides the question ahead of time, turning scientific inquiry into an instrument of confirmation bias. The EEA is fundamentally a *hypothesis-generating heuristic*. If we evolved in a certain environment, departures from it *might* result in biological dysfunction. Of course, this is not always the case. Departures from the EEA can be neutral or even lead to improvements in biological function. To find out the truth we must test the hypothesis dispassionately, without a dogmatic prior investment in the vindication of nature's perfect design.

When it comes to human biological systems, especially those that involve the biology of women and childbirth, evolutionary hypotheses can feel dangerous because of their historical association with bias. As the anthropologist and primatologist Sarah Blaffer Hrdy explains, "The early literature on the biology of motherhood was built on patriarchal assumptions introduced by an earlier generation of moralists," and "what was essentially wishful thinking on their part was substituted for objective observation."[35] Biology was the raw material for "just-so." But, she argues, the reaction against biology has gone too far in the opposite direction.

> Feminists, social historians, and philosophers were already convinced that they knew what evolutionists had to offer, that it was necessarily flawed, determinist, and uninsightful. Natural selection, and with it the most powerful and comprehensive theory for understanding the basic natures of mothers

and infants, was rejected, as social scientists and feminists took another route. That path, which led away from science, led them to reject biology altogether and construct alternative origin stories, their own versions of wishful thinking about socially constructed men and women.[36]

The answer is to abandon wishful thinking and mythic origin stories altogether. Scientists can use the EEA to generate hypotheses without romanticizing it, splitting the difference between worshiping nature and ignoring it entirely. There's no better example of how to practice this balancing act than Karen Rosenberg, who herself gave birth by caesarean, but advocates for fewer interventions in developed nations. "In many countries there are too many C-sections, like Brazil, where it's 80 percent," she said to me. "But some of the world doesn't have enough. And so I agree with what some of the natural birth people might say, but I wouldn't give the same reasons. I wouldn't say a vaginal birth is better because it's natural."

Instead of faith in nature's goodness, Rosenberg's support of vaginal birth for low-risk pregnancies depends on studies. These studies look at the potential effects of C-sections on the mother's survival rate and recovery time, as well as on the baby's microbiome, respiratory health, and a host of other outcomes. The hypothesis in these studies is that vaginal birth—obviously the default condition in the EEA—might have benefits over C-sections. Whether that's true, and what the benefits are, must wait on good evidence and could be overturned by future scientific inquiry. It's also conceivable that improvements in C-sections could eventually mitigate some of the advantages enjoyed by vaginal birth right now. The EEA can't decide anything in advance.

Similarly, Melvin Konner sees the practices of hunter-gatherers—those humans who, in theory, remain closest to the EEA—as another hypothesis-generating heuristic, one that needn't be rejected on the basis of its origins in colonial dehumanization or "noble savage" myths. His own advocacy of nutritional patterns that are more like hunter-gatherer diets, his support for breastfeeding and skin-to-skin contact between newborns and their mothers—they have nothing to do with their being "natural," and everything to do with his understanding of the evidence. That's why he has no difficulty endorsing those practices while simultaneously believing, as he said to me, that "it's probably not a good idea to give birth any significant distance from a surgical suite."

It's possible to disagree with Konner and Rosenberg about their interpretation of the evidence, but not with their undogmatic approach. Nature should generate scientific hypotheses, not foregone conclusions. Each case must be considered on its own merits. Consider birth position: hunter-gatherer women usually give birth in squatting or kneeling positions, and upright births are typical throughout the historical record up until the advent of modern obstetrics. Hearing this, a romantic might leap to the conclusion that upright births are therefore better in all respects, and giving birth on one's back serves only to make life easier for physicians. However, a 2017 meta-analysis of studies on upright birth had mixed results. While upright posture was associated with "a very small reduction [six minutes] in the duration of second stage labor" and "reduction in episiotomy rates," it also led to "increased risk of blood loss" and "increased risk of second degree tears."[37] In other words, the issue is complicated, and pretending otherwise substitutes ideology for evidence.

When natural birth advocates treat nature as God, natural birth becomes the best way—the only way—to make birth sacred and safe. Everything else is profane, meaningless, a failure. This perspective is a destructive form of fundamentalism. It confuses etymology with reality, turning "unnatural" into "unborn."

But we can supplement that traditional myth with new myths, with different etymologies. The patron saint of childbirth, Raymond Nonnatus, received his name for having been cut from his mother's dying body, reminding us that an unnatural birth, nonnatus—"not-born"—can be just as sacred as a natural birth. Had the saint been born today, an unnatural birth could have saved his mother's life.

In my father's case it was the ritual of cutting the umbilical cord that helped sacralize my birth. Whenever he tells how it happened he is always bursting with joy and pride, connected by that story to my origin, as the cord once connected me to my mother. In seeing this act as significant, he participates in a lineage of meaning-makers for whom naturalness is not the primary source of birth's meaning: the Oraibi Hopi of Arizona cut the umbilical cord with an arrow if the child is a boy, and with a grain-piling tool if it is a girl; in Punjab, India, the cord is cut with a knife belonging to an elder man if it is a boy, and with a spindle if it is a girl.[38] Arrows, tools, knives, spindles, and, for my dad, scissors. None are natural, but they can create just as much meaning as giving birth in an Australian rainforest, or your own home.

The problem is that nature, like humans, exists simultaneously as an object of scientific inquiry and a source of mythic significance. The values of these realms are articulated in drastically different terms: *perineal tear frequency* and *neonatal mortality* versus *feeling connected to womankind* and *participating in primal mysteries*. When we insist on expressing one in terms of the other, distortion is inevitable. If you favor the scientific side, you will pretend that significance is fully reducible to medical metrics, that the quality of a birth should be measured solely by how safe it is. If you favor the natural side, you will pretend that value is reducible to what's natural, that the EEA really was Eden and animal births are safe and painless. In one case, safety subsumes significance; in the other, significance is assumed to indicate safety. Both fail to capture the complexity of what it is to be human—an animal capable of artifice, a metaphysical chimera born to the impossible task of making itself whole, a survivor and also a storyteller.

The increasing dominance of scientific discourse, accompanied by the dismissal of anything that looks like a religious justification, has created a sort of rhetorical monoculture in which appeals to fuzzy factors such as "meaning" or "sacredness" don't count as objective descriptions of reality. An appeal to mythic significance becomes a confession that you believe in something false. But remember: myths need not be false—and does it always matter if they are? Borrowed from the Greek, mythos (μῦθος) means speech, narrative, fiction, plot. We live in a story-shaped world.[39] Our stories are who we are. *My father is the one who cut my umbilical cord. I am the father who was present at my daughter's birth.*

Nature gave birth to humanity. That's a true story, a true myth. If women and families feel called to participate in that myth through the ritual of natural birth, there should be space in our discussions to honor their decision on its own terms. We will disagree about the value of natural birth and what it even means, just as we will disagree on the value of feeling liberated by formula feeding and the meaning of freedom. Nevertheless, if we recognize that value, it will allow for better dialogue—honest, accurate dialogue—about who we are and what we want, struggling to stay alive and make meaning, connected by a *sipapu* to the story-shaped world that bore us.

CHAPTER 2

The True Vine

There is at least three times as much vanilla consumed as all other flavors together, and in all probability the consumer knows less of its origin than of any other material from which extracts are made, few being familiar even with the matured vanilla bean.

— E. M. CHACE, assistant chief, Division of Foods
at the USDA's Bureau of Chemistry, 1908

I am the true vine, and my Father is the vinegrower. He removes every branch in me that bears no fruit. Every branch that bears fruit he prunes to make it bear more fruit.

—JOHN 15:1–5

TO SEE THE FUTURE OF NATURAL VANILLA I had to wear a disposable lab coat, wispy and white like a hospital gown. It was meant to prevent me from contaminating a Dutch research facility, where, along with his small team, the scientist Filip van Noort is growing an astonishing array of plants, everything from garden standards like tomatoes, cucumbers, bell peppers, and tulips (of course!) to papayas, peppercorns, hops, jackfruit, and the beautiful vanilla orchids I had come to see.

Operated under the auspices of Holland's celebrated Wageningen University (mission: "To explore the potential of nature to improve the quality of life"), the facility is an imposing glass superstructure divided into separate greenhouse units, each uniquely calibrated. As I walked from one to the next it was as if I were traveling the globe: some were warm and humid like the tropics, others dry and lined with lights that supplement the Dutch sun. In a few of them, bumblebees droned about their business, imported laborers that I was told are invaluable to the operation.

Occasionally there was a unit with no plants at all, just metal vats, or rows of test tubes, or a giant steel machine festooned with dials. One was taken up by a maze of clear tubes whooshing with red liquid. It reminded me of Willy Wonka's chocolate room, but less whimsical; van Noort explained that he's growing red algae, destined to become a colorant in feed pellets for farmed salmon. Customers prefer the pink hue of wild-caught fish flesh to the otherwise naturally gray farm variety, and they prefer their pink-ifying astaxanthin from natural sources like algae rather than the petrochemicals from which it is usually synthesized. (Though there's no effect on flavor, the difference in color commands a substantial price premium.)[1]

Van Noort is in his mid-fifties, a tall, wiry man with close-cropped hair who obviously takes immense pride in the whole operation. He stopped repeatedly as we wandered the complex, enthusing about the benefits of growing plants on plastic-wrapped cubes of something called rockwool— "much easier for me to control than natural substrate"—or dilating on the intricate manipulations required to make hops flourish indoors, year-round. He's particularly proud of his operation's efficiency, citing figures for various vegetables that put him at four times the production of a typical outdoor farm, using far less land and only one-hundredth the water.

In one of the plant-free units I noticed an exceedingly complicated as-semblage of pipes and control panels, next to which hung a large banner: "Agrozone: Natural Clear with Natural Innovations." It was at this point I felt compelled to ask van Noort if what he's doing would qualify as natural.

"It's not natural at all!" he exclaimed, then paused, as if realizing what that entailed. We finally arrived at the vanilla orchid greenhouse, and both of us looked in silence at the perfect rows of vines climbing from their indi-vidual plastic pots toward the glass ceiling, his initial response still hanging in the air. Could this really be the answer to consumers' demand for more natural vanilla, a place as sterile and controlled as a surgical suite, entirely dependent on cutting-edge science and state-of-the-art technology?

"Well," he continued thoughtfully, "I use all the tools we discovered from nature, light, humidity, nutrients, so I don't think I am doing unnatu-ral things. The cannabis growers"—and here he gestured in what I assume was the direction of Amsterdam—"they think they are God, because they control everything, even the light. We use natural light and artificial light." Again he paused, and smiled. "But we are also more or less in control of everything."

Van Noort is a devout Christian, and I pressed him about the implications of playing God. "God made us with brains, to discover," he said without hesitation. "I discover how to grow plants. Everything I do is about balance, balancing the lights, the nutrients, the water. Half of my work is having—what do you call it—a green finger. We collect data, but I have to spend time with the crop. Just data would be a disaster."

Yet as our conversation went on, it became clear that van Noort had mixed feelings about technology, at least as it is currently developed and deployed. He was skeptical of genetically modified crops, not for health reasons, but because he felt that in the hands of large corporations they could end up circumscribing the autonomy of farmers who depend on them. He seemed wary of the business model that drives agricultural research more generally, including his own. Nevertheless, it was also clear that he saw the public as complicit. Their demand for natural vanilla wasn't a solution— it was a way to avoid dealing with underlying systemic problems.

"People do *this*," he said, bringing his hands up to cover his eyes. "They know that cacao is harvested by children, but they don't want to think about it. Even natural vanilla, it is easier for women and children with smaller hands to pollinate. They want natural, but they don't like knowing where it comes from."

A part of me agrees with van Noort. Many of us, myself included, are hypocrites. We say we prefer natural food but have no qualms about eating strawberries and papayas all year round. We suspect that much of what we eat is the product of unsustainable agriculture and immoral labor practices, but we eat it anyway. We cover our eyes.

However, it's not entirely our fault. The opacity and complexity of our food system make it difficult to eat well, according to anyone's criteria of what that means. For the majority of people I've spoken with, eating what's "natural"—or what bears that label—is a genuinely good faith effort to do the right thing. "Natural" is an earnest metaphor for goodness. It's shorthand for a creation myth about how food was produced and where it came from, which certifies certain foods as virtuous in every sense of the word. And if people do not see the distance between the myth of natural food and reality, perhaps it isn't willful self-deception, but rather that they've never been told a different story.

To understand vanilla, say the indigenous farmers of Papantla, Mexico, you should know that it was born from soil soaked in the blood of forbidden love.

Long ago, goes the tale, the ancient king Teniztli had a daughter of exceptional beauty: the princess Tzacopontziza. Her perfection made him proud, but also worried. What man, he thought, is worthy of such a treasure? After reflecting, he decided that no such suitor could exist and ordered his daughter taken to a nearby mountain temple, where she would be consecrated to the goddess Tonacayohua and live out her life as a virgin priestess.

For a time, Tzacopontziza accepted her role. But one day, out of nowhere, there appeared near the temple a young man named Xcatan-oxga, who had seen the princess from afar and fallen in love with her. He approached the daughter of Teniztli, and it took only moments for them to decide they would leave the temple together. When the pair tried to escape, the goddess Tonacayohua's priests stopped them. Sacrilege, they cried! The young couple was beheaded, their hearts removed, and their bodies, along with the rest of them, tossed into a ravine.

In the ravine, strange things began to happen. First all the plants dried up. Then a bush began to grow, just one, filled with leaves. Next to it grew an orchid, which within days had wound itself around the bush in a delicate embrace. When both had grown as tall as they could, the cream-colored orchid flowers turned into dark pods that gave off an indescribably exquisite aroma. The priests claimed that their goddess had made this sacred plant from the hearts of the lovers. They called it *xanath*—and to this day, that is the Papantlan word for vanilla.[2]

The indigenous story of vanilla's creation is a cautionary tale about undue attachment to religious purity, and the necessity of change over time. The king wants to preserve his daughter in her original, virginal form, but she refuses. Together with her lover she breaks from the previous generation. Vanilla was born of human passion, its fragrance a reward for abandoning traditional norms.

In what it emphasizes, this myth is remarkably different from the myth of natural food. When "natural" is used to describe a food like vanilla, it assures us that the food has been around for a very long time, in more or less the same form. Stasis, purity, unchanging essence—these are at the heart of the natural myth. "Natural" means the *original* version—and, as with holy books, "original" means best. Natural products are *authentic*—another word of praise that etymologically assumes the virtue of being close to something's author. There's no chance that supermarket shoppers will mistake "all natural" for a warning instead of an endorsement, because the origin story behind the word tells them not to.

The natural myth is so compelling that industry titans are harnessing its power by reformulating iconic products. Many of these products depend on the world's most popular flavor: vanilla. In the summer of 2017, McDonald's announced that its vanilla soft-serve would no longer contain artificial vanilla flavoring. McFlurries and McCafé Shakes are now made exclusively with "natural flavors"—just like Hershey's Kisses, which switched over from synthetic vanillin in 2015. (Both have also eliminated high-fructose corn syrup in favor of more "natural" sugar.) "We've been raising the bar at McDonald's on serving delicious food that our customers can feel good about eating," said Darci Forrest, senior director of menu innovation. "Soft serve now joins the other changes we have made."[3]

She's right. Customers do feel good about eating natural McCafé Shakes and Hershey's Kisses, in large part because natural has come to be synonymous with whatever positive values we hold: environmental sustainability, flavor, and, above all else, health. A sampling of representative headlines captures the reaction that companies hope to inspire in consumers with their changed formulations:

> *"We're Lovin' It! McDonald's Cuts Out the Bad Stuff in Its Soft Serve"*
>
> —POPSUGAR

> *"Healthier Ice Cream? McDonald's Removes Artificial Flavor from Vanilla Soft-Serve"*
>
> —DAILY MAIL

> *"Hershey to Offer Healthier Kisses This Holiday Season"*
>
> —REUTERS

> *"Healthy and Delicious: Hershey's Will Switch to Chocolate with Simple Ingredients"*
>
> —E! ONLINE

Wisely, the companies themselves don't advertise their products as healthier; in the case of vanilla, virtually no scientific evidence supports the claim. Fortunately for them, there's no need to make it. Consumers do so themselves because it's right there in the origin myth. Nature is perfect, intentional, and original—therefore the best foods are those closest to how nature made them. Natural ingredients are better, says the myth, healthier and tastier, not to mention easier on the environment.

The promise of natural foods is a return to a simpler, earlier, more authentic time, when people enjoyed what nature authored, not the products of an artificial lab. We feel safer with what we can understand, and we feel like we understand where natural products come from. It's simple—they just grow out of the ground, and we cook them using the traditional methods that humans used for centuries until modern technology corrupted preparation methods with microwaves and chemicals. Hershey's website features "The Simple Promise"; McDonald's food philosophy is "The Simpler the Better"—invoking a better, less-complicated culinary time when sugar was sugar, vanilla was made from beans, and people didn't have to worry about their food.

"Don't eat anything your great-grandmother wouldn't recognize as food," wrote Michael Pollan in his influential bestseller *In Defense of Food*, advice meant to steer consumers toward traditional preparation techniques and away from artificial flavors, emulsifiers, stabilizers, preservatives, and sweeteners.[4] Why? Because these unnatural innovations distance us from a better culinary time. Unnatural food is "bad stuff," as it says in the *PopSugar* headline—bad for the world and bad for us. Natural products, by contrast, are sourced directly from some not-so-distant Eden, when humans lived in harmony with the earth and died from old age, not cancer or complications from type II diabetes.

But creation myths can be misleading if we aren't careful. Vanilla orchids did not originally grow from the blood of a slaughtered princess, and it would be a mistake to base purchasing decisions on the literal truth of that belief. Likewise, natural products rarely come from where we think they did. Not only that, but even if ancient provenance can be proved, it is no guarantee of quality. The violation of old customs can produce great triumphs. Just ask the mythical priests who first smelled vanilla pods in the ravine near their temple.

What you are about to read is a history of vanilla, but it is also a test of the natural myth. Was there really a simple, authentic world of natural ingredients that got progressively worse over time, as people disturbed their natural state and departed from traditional preparations?

Here is what botanists and anthropologists know for certain.[5]

Since the first appearance of vanilloid orchids on earth, around sixty million years ago, no animal but humans has ventured to eat their fruit, since, like many of our foods, it is basically inedible without significant processing. Before we came along, the tangled vines of *Vanilla planifolia* snaked through prehistoric Mesoamerican rainforests, climbing support trees toward the canopy. As the hot tropical sun arced through the sky, their fragile flowers slowly opened, pollen-heavy stamens waiting for a visit from tiny, native euglossine bees, their only known natural pollinator.

Most of the flowers go neglected, withering by early afternoon and drifting down to the thickly packed leaves below. The lucky ones take another six months to produce slender green pods, which remain scentless until fully ripe, at which point they split open and fall to the forest floor. The powerful scent likely attracted the attention of preagricultural hunter-gatherers who once roamed those forests. We know very little about the role of vanilla in their lives, but it appears they may have harvested ripe beans to perfume themselves, and used the liquid from the green ones to treat infections and venomous bites. These are the only known uses of vanilla beans taken directly from nature and left in their natural form.

Today, my local coffee shop happily provides free shots of vanilla syrup to anyone who asks. Eighteen thousand metric tons of vanilla flavor are sold every year. Although most of it is used for manufacturing ice cream, chocolate, soda, baked goods, and other food products, a significant amount winds up in shampoo, soap, shaving cream, perfume, detergent, housecleaning products, toothpaste, sunscreen, insect repellant, and pharmaceuticals. That subtle flavor in glossy pill coatings? Vanilla. The scent of your favorite fabric softener? Vanilla. Even animals like it: studies show that piglets fatten up more quickly on vanilla-flavored fodder.

From rare Mesoamerican orchid to livestock feed. How did it happen?

In short: unnaturally. The first people to change humanity's relationship with vanilla were early Mayan forest gardeners. No one knows for certain, but it's possible that these sophisticated silviculturists collected vine cuttings from the forest and placed them in domestic gardens, concentrated collections of fragrant, beautiful plants that marked ritually important spaces. Another possibility is that cacao farmers discovered wild orchids growing in their unnatural orchards and decided to turn them into a second commercial crop. However it happened, history was made. Humans had launched vanilla on its long journey from prehistoric rainforests into Starbucks lattes.

In addition to domesticating vanilla, the Mayans and later Mesoamerican peoples developed invaluable processing techniques. The beans were picked green and subjected to an elaborate four-stage curing procedure: killing, sweating, drying, and conditioning. "Killing" artificially disrupts enzymatic processes. If you've ever cooked with vanilla beans, they were probably killed with freezing, oven-heating, or hot-water scalding. These efficient methods were unavailable to the forefathers of vanilla. They killed the beans by placing them under dark fabric in direct sunlight for several days. Dead, the beans were then tied into bunches and tightly wrapped in cloth, "sweating" out moisture at an even rate for another several days. At this point they were ready to be dried, a three-month affair that involved laying them out on racks in the gentler morning sun, then transferring them to the shade every afternoon. Finally, the beans were placed in containers for five to six months of "conditioning," the last stage before they could be used to manufacture medicines, flavor beverages, and scent tobacco.

Meanwhile, a different set of technological innovations—nautical, not agricultural—allowed emissaries of the Spanish Empire to visit the so-called New World. Upon encountering vanilla, the Spaniards immediately began importing it. Disappointingly, the orchids they brought home with the pods rarely flowered and never fruited, despite the best efforts of European botanists. A mere two hundred years ago, vanilla orchids still grew exclusively in their native habitat, the vines carefully cultivated by indigenous farmers, their ghostly white flowers pollinated by a rare species of bee. Throughout the rest of the world, the scent and flavor of vanilla were known only to elites: the wealthy few who could afford high-end perfumes, tobacco, and confections laced with the New World luxury.

Then, in 1841, a twelve-year-old French slave named Edmond Albius discovered how to artificially inseminate the orchid flowers. No longer dependent on natural pollinators, commercial orchid farms began to flourish outside Mexico. Production went up, and so did demand. But Albius's method was extremely labor intensive. Recall that vanilla flowers only open for a few hours. During this small window of time you must carefully split the flower with a thin stick, lift a membrane called the "rostellum," which prevents self-pollination, and gently press the tiny anther to the stigma. Over a century later, we have not improved on this method. Every single bean on the market owes its existence to the painstaking hand-pollination of an individual flower. (Even van Noort's team uses this method, and on the day of my visit one of his assistants complained of having had to pollinate 480 flowers.) This is why vanilla is a hundred times more expensive by weight than coffee—though, to be fair, it is currently worth a little less than silver, which was the original currency paid for it by Aztec nobility.

To make vanilla genuinely affordable, scientists had to bypass orchids entirely, unlocking vanilla's chemical secret and duplicating it in a lab. In the late nineteenth century they succeeded, first by isolating vanillin molecules, then by discovering how to synthesize vanillin from clove oil through a process known as isomerization. Vanilla steadily increased in popularity, and companies developed multiple synthetic production routes. Pine sap, clove oil, bacteria, genetically modified yeast, even cow dung—the fragrant vanillin molecule has been manufactured from all of these and more. Today, an estimated 85 percent of vanilla flavor is derived from petrochemicals.

A remarkably small amount of vanilla flavor, under 1 percent, actually comes from vanilla plants, and all but a tiny fraction of that 1 percent is grown in nonnative habitats like Madagascar and Indonesia. There the orchids are planted, pollinated, tended, and protected from rampant theft by impoverished farmers, whose monthly salary would barely pay for a pound of the beans that get shipped, flown, and driven from one point of processing to the next, in vehicles that run on the same petrochemicals used to synthesize artificial vanillin, destined for the distant kitchens of chefs and bakers who will process them still further.

It seems reasonable to distinguish natural from unnatural by drawing the line between vanillin synthesized in laboratories and vanilla that comes from orchids. But the truth is that with natural, the line is always relative to our contemporary perspective. Every innovation in the history of vanilla would have been highly unnatural at the time, the pinnacle of technological progress. Ships look natural from the windows of airplanes; vanilla pods seem natural to the McFlurry drinker. But when they were first developed, ships and cured vanilla pods represented radical new forms of travel and food production. Going to the New World was like going to Mars, and eating farmed vanilla was once as strange as frying up a lab-grown burger would be today.

If every generation had followed Michael Pollan's advice to eat like their great-grandparents, we wouldn't have vanilla at all. But the advice feels sound because it draws on the creation story that's contained in the idea of eating naturally. Through the alchemy of the story, mythic time becomes real time; the residents of paradise are no longer Adam and Eve but our relatives from a few generations ago. Paradise as historical reality is attainable simply through changing our diet.

However, a utopian vision of our culinary ancestors does not square with historical reality—even if utopia is just a place where synthetic additives do not exist and food is better for it. And, as the saying goes, those who do not understand history are doomed to repeat it. Pragmatic, realistic improvement of our diets will not happen if we model our solutions on an invented natural paradise, one which does not stand up to a clear-eyed examination of the past, and, in any case, could never sustainably satisfy the appetites of over seven billion humans, at least if they want to occasionally experience the taste of vanilla.

My own great-grandmother would have been making food in 1906, the year that the US Congress passed the Pure Food and Drug Act. The food portion of that act, which drew its language from an even earlier 1881 New York food law, was enacted in response to rampant problems with dangerous and fraudulent adulteration. Unnatural Hershey's Kisses made with artificial vanillin and high-fructose corn syrup look much safer when compared with nineteenth-century penny candies. These, according to an 1833 article in the medical journal *Lancet*, were routinely made with "red oxide of lead, chromate of lead, and red sulphuret of mercury," then wrapped in papers printed with poisonous dyes known to inflame the gums of children who sucked them.[6] (Nor were the problems restricted to cheap treats: a 1901 trade journal describes how powdered glass gave expensive French confections "a glittering.")[7]

The fifth case brought under the 1906 food law was a lawsuit against C. B. Woodworth Sons Company for the misbranding of vanilla extract. According to the USDA's Bureau of Chemistry, the "Double Extract of Vanilla" contained "a mere trace of vanilla and a coal-tar dye to impart the color of pure extract."[8] It's likely my great-grandmother would have used just such a product in her baking, since the real thing was substantially more expensive.

The need for regulatory standards demonstrates the prevalence of imitation and artificial flavors, which were widespread by the mid-nineteenth century. In his massively successful book *Information for Everybody: An Invaluable*

Collection of about Eight Hundred Practical Recipes, Dr. Alvin W. Chase detailed innumerable methods of cutting costs by replicating otherwise-expensive ingredients. Throughout, Chase insisted on the equivalence of his formulas to their natural counterparts: "This is called the 'Saturated Tincture'; and use sufficient of this tincture to give the desired or natural taste of the raspberry, from which it cannot be distinguished." Flavor and syrup recipes from the time regularly called for "natural" and artificial ingredients, combining fruit juices with such chemicals as acetic ether (strawberry flavor), butyric ether (pineapple flavor), and, as soon as it was readily available, artificial vanillin.[9]

One notable exception was Coca-Cola, formulated in 1886 by the ex-Confederate general and morphine addict John Pemberton, who was looking for an alternative to his current drug of choice.[10] The original recipe included genuine vanilla extract, along with two other natural products: kola nuts, for the flavor and caffeine, and coca leaves, for the cocaine. The "cocaine kick," common to patent medicines of the day, was touted by Pemberton as "a most wonderful invigorator of the sexual organs."[11]

The drink surged in popularity. Coca-Cola delivery trucks were nicknamed "dope wagons." And when concerned members of the temperance movement complained, Asa Candler, a businessman who had taken over from Pemberton, was incensed: "Do you really want us to change Coca-Cola, the purest, most healthful drink the world has ever seen?"[12]

Continued public outcry led to a cocaine-free version, the original New Coke. And although it is made with highly processed coca leaves—yes, even now, a New Jersey chemical processing plant extracts the psychoactive components with a special technique invented in 1929—modern Coca-Cola's formulation must be counted a marginal improvement, at least in terms of healthfulness, over the natural, pure, traditional version that Pemberton and Candler once defended.[13] (The clearest modern turn for the worse is portion size, which, needless to say, has nothing to do with naturalness.)

What's true of products is also true of production methods. Whatever their defects, much-maligned modern sweeteners such as high-fructose corn syrup and aspartame are marked improvements over *defrutum* (also known as *sapa*, depending on the concentration), an ancient food additive so toxic that it has been implicated in the fall of the Roman Empire. Centuries before the Aztecs learned to flavor cacao beans with vanilla, Roman chefs were boiling grape must down into this thick syrup used to sweeten and preserve sauces, delicacies, and wine—the last of which was drunk by nobility at the astonishing rate of two liters per day. Unfortunately, two thousand years ago there were far fewer materials available for the manufacture of cooking and storage vessels. Lead, being one of the most easily accessible naturally occurring elements, was a popular choice, employed not only in pipes—the word "plumbing" comes from the Latin *plumbum*, for "lead"—but also pots and cauldrons. "Preference should be given to lead vessels . . . in boiling *defrutum* and *sapa*," wrote Pliny the Elder in a classic treatment of the process.[14] As a result, the syrup contained high levels of lead acetate, which had a sweet taste and also acted as a fungicide. Gourmet chefs from as far back as 150 BCE offered creative suggestions for how to use the leaded treat, from enhancing wine to preserving quinces, including this recipe for flamingo found in *Apicius*, the world's oldest surviving cookbook: "Roast the bird. Crush pepper, lovage, celery seed, defrutum, parsley, mint, shallots, dates, honey, wine, broth, vinegar, oil, reduced grape must to taste."[15]

Experts disagree about exactly how much lead exposure would result from a traditional Roman diet. Geochemist Jerome O. Nriagu's 1983 book, *Lead and Lead Poisoning in Antiquity*, estimates that just a teaspoon of daily *defrutum* was enough to cause chronic lead poisoning, and suggests that upper-class Romans consumed at least 250 micrograms per day of lead, twenty times higher than the US FDA's maximum allowable daily intake of 12.5 micrograms. On this disastrous diet, argues Nriagu, emperors and aristocrats became goutish and brain-damaged, potentially to the point where they could no longer effectively govern or manage their personal lives. Others have accused him of hyperbole—though the debate is not settled—but it remains true that whatever the amount, Roman lead levels would be intolerable according to today's safety standards. Roasted flamingo with *defrutum* and a goblet of wine? You'd be better off sticking to fried chicken and an artificially flavored Coca-Cola Vanilla, or even better, an (unleaded) glass of chlorinated, fluoridated tap water, safely piped into your home via untraditional plumbing materials like galvanized steel, polybutylene, and polyvinyl chloride.

The history of beverage manufacturing fits with the broader history of culinary art, which is essentially a parade of unnatural manipulations, starting with the invention of cooking itself. Heating, cooling, mixing, aging, curing, drying, preserving, fermenting, distilling, purifying, pasteurizing, homogenizing: all of these are decidedly human innovations, technologies that only seem traditional and natural in hindsight. That's why it's hard to see natural purity in the consumption of vanilla, whether artificially synthesized

or picked off artificially inseminated orchids. Be it Montezuma's legend-
ary fifty daily cups of *xocolatl* (the unsweetened ancestor of hot chocolate);
high-end natural vanilla flavor, prepared by soaking beans in ethyl alcohol
and propylene glycol; or vanillin produced through processing petroleum—
some methods may be older than others, but all involve the hand of man.
Pliny himself writes of *sapa* that it is made "by Art, not Nature," and the
same can be said of wine, *xocolatl*, and vanilla in any consumable form.

Granting that naturalness exists on a spectrum—Aztec vanilla beans are
surely closer to nature than vanillin made from steam-cracked petrochemical
feedstock—vanilla and Roman wine demonstrate that "more natural" cannot be reliably indexed to safety or healthfulness. As the British food writer
Bee Wilson points out in *Swindled: The Dark History of Food Fraud*, there are
no records of anyone getting vanillin poisoning (though some farm workers
have been stricken by vanillism, a severe allergic reaction to handling vanilla
beans). Wilson compares the allowable UK guidelines for maximum vanillin content—20,000 mg/kilo—with those for piperine—1 mg/kilo—which
occurs naturally in black pepper. "It would be hard to make the case that
vanillin poses health risks," says Wilson.[16]

She's right. Vanillin just isn't a health risk. Far more dangerous are the
excessive discretionary calories, natural or otherwise, that come in the form
of McFlurries and Hershey's chocolate—especially if the "natural" label lib-
erates people to consume them in excess. Similarly, the riskiness of *defrutum* may itself be eclipsed by the danger of consuming the *two liters of wine
per day* believed to be the daily average of Roman aristocrats. (The alcohol
levels were lower than that of most modern wine, but consumption still far
exceeded standard modern guidelines.)

Should we really prefer natural, traditional drinks? Alcoholic beverages
have the distinction of being the most traditional drink next to water. Ber-
ries ferment in nature, and some anthropologists credit the invention of agriculture to humans' desire for alcohol. Yet this natural beverage is highly
addictive, the current fourth-leading cause of preventable death worldwide.
A drink or two a day may be good for your health—but many humans have a
natural tendency to drink more, with devastating consequences.

As for the second-leading cause of preventable death, just after hyper-
tension? It's tobacco, a plant that has wreaked untold havoc across the globe
ever since Spaniards brought it back from the New World, along with the
vanilla beans traditionally used to flavor it.

In addition to distorting our perception of food safety, the myth of natural food also distorts our sense of taste. The debate over vanilla flavor dates back to those late nineteenth-century recipe books that offered cheap substitutions for thrifty home cooks. Combing through them, the food historian Nadia Berenstein discovered multiple recipes for "Extract of Vanilla—Without Vanilla." Some recipes were composed entirely of natural products, creative concoctions of prunes, molasses, botanical resins, and, most importantly, tonka beans.[17] Shriveled, squat, and shorter than a vanilla bean, the tonka bean was a flavor miracle thanks to its high concentration of the chemical coumarin. Concern about chronic coumarin exposure causing liver toxicity in animals led the FDA to make tonka beans illegal. They remained tasty, however, and despite occasionally being busted by the Feds, gourmet chefs in America still smuggle them in: "As long as you don't use a copious amount of it—obviously a copious amount could cause death—it really is delicious," said Thomas Raquel, head pastry chef of New York's Le Bernardin, to the BBC.[18]

Natural and artificial coumarin were cornerstones of early vanilla substitutes, and manufacturers insisted that consumers couldn't tell the difference between fake stuff and the real thing. Then, in 1875, French scientists successfully synthesized vanillin from clove oil, and within two decades the controversy over naturalness and flavor went mainstream. By this time the general public had become accustomed to vanilla extract enhanced with tonka beans, coumarin, or a jolt of synthetic vanillin, and Berenstein describes how manufacturers objected strenuously to new federal rules that prohibited such adulteration. People loved their version, not the natural one! "When so-called Vanilla Extract is fortified by Vanillin," wrote the San Francisco pharmacist A. L. Lengfeld in 1905, "it is preferred by a majority. Forbidding [artificial additives] seems like catering to the prejudice of the ignorant against any food product that is not NATURAL in its origin." Others pointed out that adding tonka extract to vanilla extract made for better retention of vanilla flavor "in the process of baking or where heat is applied."[19]

Is it true that people prefer adulterated vanilla and fake vanillin over the real thing? Early twentieth-century manufacturers weren't performing double-blind taste tests, but modern culinary enthusiasts certainly have.

Time and time again, participants have been shocked by the results. *Cooks Illustrated* even provided this preamble to a 2009 taste test: "In two past tastings of vanilla extract, we reached a conclusion that still amazes us: It matters not a whit whether you use real or imitation vanilla, because you can't tell the difference when you bake. But at a recent editorial meeting, we took a poll: Did that mean anyone had stopped buying the real thing? No. Our test cooks believed firmly that natural vanilla is the best choice. So we returned to the test kitchen for a definitive tasting."[20] The third time around didn't change much. When the extract remained uncooked—as in eggnog or added at the end of a pudding recipe—real vanilla proved superior to artificial vanilla. But in baked goods like cookies and cakes? "We have to admit," read the report, "there's not much difference between a well-made synthetic vanilla and the real thing."

Others have come to similar conclusions. In a blind taste test with cookies, the food blog *Epicurious* found that people either couldn't tell the difference or preferred cookies made with imitation vanilla. Tasters at *Serious Eats*, another food blog, had difficulty distinguishing between the two in everything but eggnog—and actually exhibited a strong preference for ice cream made with the fake stuff. Those early manufacturers were right. Although naturally occurring compounds in vanilla beans give genuine extract a more complex flavor profile, the compounds cook off at high temperatures, rendering their presence undetectable even to professional tasters. In the words of acclaimed chef and food writer J. Kenji López-Alt, "If you use vanilla regularly in baked goods like cookies and cakes, there's no reason to spring for the fancy stuff, or even the real stuff—artificial extract will do just fine."[21]

Vanilla counterfeiters have long known that it's difficult to distinguish between artificial and natural vanilla. For nearly a century they've been passing off fake stuff as the real thing, or spiking low-quality extracts with synthetic vanillin. This has led to an escalating arms race between criminal chemists and frantic regulators who employ an array of techniques—gas chromatography, liquid chromatography, mass spectrometry, and stable ratio isotope analysis—to detect forgeries. Writing in the *New Yorker*, Roald Hoffmann, winner of the 1981 Nobel Prize in chemistry, describes one remarkable episode in the race when forgers began adding tiny amounts of radioactive ^{13}C (an isotope of carbon that can be used by scientists to establish the identity of a given substance) to their synthetic vanillin. The artificial version does not have the same $^{13}C/^{12}C$ ratio as natural vanillin, so forgers increased the amount of ^{13}C to disguise it. This approach is now out of date but the arms race continues, and sophisticated scientific detection techniques continue to struggle with the difference between natural and artificial, just like human palates do.[22]

Potential reasons for wanting to correctly identify natural vanilla include not only healthfulness and flavor, but also ethics. One primary concern in this respect is environmental impact. Artificial vanillin was once synthesized from lignin, a by-product of wood pulp processing. At first this seemed like an ideal route in terms of sustainability. Paper mills were producing waste in the form of something called acid sulfite pulping liquors, and in 1875—the same year French scientists synthesized vanilla from clove oil—mill employees reported that the liquor had a vanilla-like odor. Researchers in Canada and the United States figured out how to convert the waste product into vanillin, and by 1981 the Ontario Pulp and Paper Company supplied fully 60 percent of the world's synthetic vanillin. Voila! A renewable waste product that could be converted to good use.[23]

But there was a catch. For every kilogram of vanillin, the synthesis produced 160 kg of "caustic liquids" that had to be disposed of. In the absence of an environmentally friendly way to do so, factories were forced to close down their vanillin operations, and today only Borregaard Industries of Norway still manufactures synthetic vanillin from lignin. Now, most of it is made using a process that depends on guaiacol, a "feedstock" that occurs naturally in a variety of substances but is synthesized in industrial quantities from fossil fuels. Like vanillin, natural and synthetic guaiacol are "bio-identical," and the synthetic version is far cheaper. Although guaiacol synthesis depends on a nonrenewable source instead of trees, it results in significantly less waste. If you're eating Nutella, you're eating synthetic vanillin that was once a fossil fuel—not exactly an appetizing origin story.

Sadly, authentic vanilla beans come with ethical problems of their own, both environmental and humanitarian. Most vanilla beans are grown in Madagascar, the world's ninth-poorest country, where average income sits at just under $1,000 per year. Because of inadequate government oversight, vanilla's extraordinarily high value has put it at the center of numerous horrific practices. A 2016 report by the watchdog group Danwatch does not mince words: "You may be buying stolen vanilla cultivated by children."[24] Recalling Edmond Albius, the slave who discovered how to pollinate vanilla,

Danwatch exposes how modern vanilla farm laborers are often children who have no choice but to painstakingly tend crops destined for wealthier lands. Experts believe these children make up a staggering 32 percent of the workforce in Madagascar's orchid fields, which means the real vanilla in your Häagen-Dazs may have been handled by someone like Xidollien, a shy nine-year-old in a ragged soccer jersey photographed by Danwatch working on a farm in the Sava region of northeastern Madagascar. The sourcing of vanilla is so complicated, and so unregulated, that even conscientious companies have difficulty knowing where their vanilla comes from—and most are not conscientious. When van Noort told me that people cover their eyes when it comes to the source of their food, this is what he meant.

As if child labor weren't enough, vanilla bean production is plagued by money launderers and rampant deforestation, all of which occur as the government of Madagascar turns a blind eye. Reporting for the *Guardian* in 2018, Jonathan Watts documented how traders in illegal rosewood—a market worth hundreds of millions in Madagascar alone—invested in vanilla to launder the money they'd made.[25] As prices shot up to nearly ten times the previous amount, the crops became irresistible to thieves, who would raid farmers' property and steal the beans. Since police were bought and paid for, farmers took justice into their own hands and "hacked and stabbed [the thieves] to death with machetes and harpoons," according to one eyewitness. "I think it's good," he said. "We have our own guard now." Meanwhile, chainsaws are buzzing in Masoala National Park—a sanctuary for endangered lemurs—as locals spend an influx of big vanilla money on new homes, cutting down lumber trees at night to avoid satellite detection. And who can blame them? Their homes are likely to be far smaller than the ones occupied by those of us who can afford their crops.

None of this is an issue if you're eating something that tastes like vanilla made with "natural flavor," since "natural flavor" probably indicates there are no actual vanilla beans in your Hershey's Kiss or McFlurry. Instead, it's likely that you're consuming vanillin synthesized from what was once eugenol (extracted with solvents from plants) or ferulic acid (a component of rice bran). According to the FDA, "natural" applies to any "spice, fruit or fruit juice, vegetable or vegetable juice, edible yeast, herb, bark, bud, root, leaf or similar plant material, meat, fish, poultry, eggs, dairy products, or fermentation products thereof." The last element of the list is key: *fermentation products thereof.* Employing cutting-edge research, companies have successfully produced proprietary strains of bacteria, fungi, and yeast that can ferment natural feedstocks and produce vanillin. Since the original source material is organic and the synthesis is fermentation, the end result qualifies as natural.

Perhaps the most controversial version of this production method involves genetically engineered yeast that can convert glucose into vanillin through fermentation. Evolva, the Swiss company behind it, has attempted to anticipate objections by making a short cartoon that features a dialogue between a skeptical home baker and a friendly scientist named Eve, centered on a contested vision of "natural."[26] "What's the point of tweaking natural yeast?" asks the baker at the beginning of the cartoon. Her question summons up Eve, the avatar of a friendly scientist, who explains that fermentation has been used for centuries in bread-baking and brewing. "But brewing and fermentation are natural," insists the baker. Again, Eve points out that modern GMO technology is simply a more efficient version of the age-old process of selecting for yeast strains. She argues that vanillin made with GMO yeast is "more natural" and "more sustainable" than vanillin made from petrochemicals, and a better alternative to the deforestation that would be required to farm more vanilla orchids. To drive the point home, the video concludes with a happy lemur jumping into Eve's arms.

Not everyone is convinced by such arguments, and objections to the unnatural nature of yeast-derived vanillin continue to exert influence. In 2013, citing its commitment to "quality natural vanilla" and "small-scale vanilla farmers," Friends of the Earth sent a letter to Häagen-Dazs imploring the company to reject vanilla produced with synthetic biology, "an extreme version of genetic engineering" that could never result in anything "natural" or "sustainable." The letter objects to Evolva's dependence on "synthetic DNA not found in nature," and repeatedly emphasizes the importance of *natural* vanilla ice cream, using the word thirty times in the space of two pages. A few months later, Häagen-Dazs announced that its ice cream would not be flavored with synthetic vanilla. "It's wonderful that Häagen-Dazs has confirmed that it won't use vanilla produced using synthetic biology," said Michael Hansen, a representative of *Consumer Reports*, "since ingredients derived from synthetic biology are not natural."[27] Most likely, Häagen-Dazs was eager to avoid another public relations debacle—just a year earlier it had been sued for claiming its ice cream was all-natural despite being made with Dutch-processed cocoa, a common ingredient that's produced using potassium carbonate, a synthetic alkalizing agent.

The misleading implications of "natural flavor" went viral in a 2013 video created by the food blogger Vani Hari, aka "the Food Babe."[28] Hari warns consumers that their favorite foods might contain "beaver butt" disguised as "natural flavor," because "tons of foods" are flavored with castoreum, a secretion harvested from beavers. "It's far cheaper to use castoreum, or beaver's butt, to flavor strawberry oatmeal, than using actual strawberries," she said. "They don't want you to know what's in that natural flavor, either. Just try to call the company and ask. They'll tell you it's proprietary. But there's nothing natural about the flavors they produce." Her concluding advice conjures a vague vision of nature's intentions, and how they are violated by industrial production practices: "Humans were meant to eat real food—not fake industrial food created in a laboratory, and that's why I recommend not buying food with the ingredient 'natural flavor' on the label. Who wants to be tricked or lured into thinking a food tastes better or smells better than it should or eating beaver butt for goodness' sake?!"

As many journalists pointed out in the wake of its release, Hari's video is not entirely accurate. In the nineteenth century, beavers were hunted and killed to near extinction for pelts and scent glands, but no longer. Now castoreum is obtained by anesthetizing them and milking their castor sacs (located near the anal glands), a labor-intensive process that yields small amounts of costly white ooze. Once used to add depth to high-end flavorings, today you'd be hard pressed to find castoreum in any foods. Annual production is around 300 pounds a year, compared with over 2.5 million pounds of vanillin. The one place you will find it is a traditional Swedish spirit called Bäverhojt ("beaver shout") that's made by steeping the sacs in vodka for two weeks, a process far closer to ancient food manufacturing techniques than the industrial processes that frighten Hari.[29]

Errors aside, Hari is right to point out that the FDA's current labeling standards create an enormous gap between the origin story implied by "natural" and the reality of how natural flavors are produced, a gap that companies are happy to capitalize on. Though the natural flavors that mimic vanilla beans in Hershey's Kisses or McFlurries are unlikely to have been sourced from beaver butts, they were almost certainly produced in high-tech

Apparently, I'm not the only one who feels this way about farming for nature. "A lot of worshipers come here," said Rosa, who works in the farm's food shop and is studying to become a registered dietitian. "They want to think that all their food will be from farms like this. But it is impossible. We could never feed everybody."

It turns out that even Hoeve Biesland, with its tiny production capacity and extraordinary devotion to its mission, wasn't able to stick to the closed nutrient cycle. Instead, they had to switch to a "closed balance," where input of feed is offset by export of manure. Rosa agreed that the farm is beautiful, and it would be tragic if such places disappeared. Like me, she's willing to pay extra money for the experience of eating products that are produced there. But she has no illusions about the high prices and low yields of farming for nature, and she was adamant about the need for a multipronged approach to feeding the world. "Besides," she said, gesturing at a fridge stocked with farm-fresh delicacies, "even we aren't completely natural. The vanilla flavor in our yogurt isn't real."

Natural has always been synonymous with an ideal culinary world. For advocates of vegetarianism like Jean-Jacques Rousseau and the poet Percy Shelley, meat-eating was immoral, harmful . . . and therefore unnatural. "The indifference of children towards meat," argued Rousseau in his treatise *Émile*, "is one proof that the taste for meat is unnatural."[31] In the popular 1813 essay, "A Vindication of Natural Diet," Shelley blamed flesh-eating— "unnatural diet"—for virtually every problem of his time, from crime to disease. Later, when food processing became mainstream, the definition of unnatural changed to fit the times. "It is nearly certain that the primitive inhabitants of the earth ate their food with very little, if any artificial preparation," wrote health pioneer Sylvester Graham, of graham-cracker fame. "Food in its natural state would be the best."[32]

Thankfully, we can value what's natural on its own terms without asserting its perfection. As the chemist Roald Hoffmann observes, even the strongest skeptics of the natural/unnatural dichotomy honor some version of naturalness in their own lives: "[They] go home to houses with picture windows and not with large photographic enlargements of landscapes. In their homes grow real plants, not artful plastic and fabric imitations. They will avoid like the plague plastic shingles on their house and wood grain imitations in their dining room furniture."[33]

Why? Hoffmann—no wooly-headed romantic, but a Nobel laureate in chemistry!—speculates that part of the answer involves what he calls "spirit," a profound desire for connection to the primal order that predates humans and factories:

> I believe that our soul has an innate need for the chanced, the unique, the growing that is life. I see a fir tree trying to grow in an apparent absence of topsoil, in a cleft of a cliffside of Swedish granite near Millesgarden, and I think how it, or its offspring, will eventually split that rock. The plants trying to live in my office remind me of that tree. Even the grain in the wood of my desk, though it tells me of death, tells me of that tree. I see a baby satisfied after breast feeding, and its smile unlocks a neural path to memory of the smiles of my children when they were small, to a line of ducklings forming after their mother, to that tree.[34]

For me, eating food that's close to nature—even if it is not fully natural— evokes this satisfaction of spirit. The feeling is so powerful that it can lead us to confuse naturalness with goodness. But goodness is many things: corporate transparency, environmental health, farmers' welfare, animal welfare, affordability, flavor, healthfulness, ease of transportation, ease of preparation, cultural traditions, novelty, and, yes, naturalness. When nature is God and natural is goodness, eating what's natural magically aligns all values—embracing simple utopianism at the expense of addressing reality. In a series of striking contrasts John Steinbeck dismisses this nostalgic utopia, while recognizing his own participation in creating it:

> Even while I protest the assembly-line production of our food, our songs, our language, and eventually our souls, I know that it was a rare home that baked good bread in the old days. Mother's cooking was with rare exceptions poor, that good unpasteurized milk touched only by flies and bits of manure crawled with bacteria, the healthy old-time life was riddled with aches, sudden death from unknown causes, and that sweet local speech I mourn was the child of illiteracy and ignorance.[35]

The grave issues that confront our immensely complicated food system cannot be resolved with a division between natural and unnatural, the good old days of great-grandmother's kitchen and modernity's industrial fall from grace. We need to recognize that overuse of pesticides is bad because it harms the environment, not because it is unnatural; that confining chickens to tiny cages is bad because it is cruel, not because it is unnatural. Likewise, no matter how natural wild-caught fish might be, it is still bad to overfish them, and no matter how natural alcoholic beverages might be, they still kill more people than Diet Coke. There can be no perfectionist theology of food, not in our messy world of incommensurable values.

Better, instead, to exchange the myth of past culinary paradise for a more nuanced history, reckoning with our ambiguous relationship to nature. On this the ancient Aztecs can help. Their mother goddess was named Coatlicue, and she governed creation and destruction. In traditional depictions her breasts, emptied after nursing, rest on top of a skirt made from snakes and buckled with a skull. According to legend, she eventually sacrificed herself to create the world as we know it today, just like vanilla was born from the blood of dead lovers.

Nature is a cycle of life and death, of innovation after destruction. It is sacred and terrifying and constantly changing, just like our relationship with it. Coatlicue reminds us of this dynamism—a reminder that's especially necessary when we are repeatedly told that nature is not a process, but a state.

CHAPTER 3

States of Nature

I am as free as Nature first made man,
'Ere the base Laws of Servitude began,
When wild in woods the noble Savage ran.

—JOHN DRYDEN, *The Conquest of Granada,* 1670

In such condition, there is no place for Industry; because the
fruit thereof is uncertain; and consequently no Culture of the
Earth; no Navigation, nor use of the commodities that may
be imported by Sea; no commodious Building; no Instruments
of moving, and removing such things as require much force;
no Knowledge of the face of the Earth; no account of Time; no
Arts; no Letters; no Society; and which is worst of all, continu-
all feare, and danger of violent death; And the life of man,
solitary, poore, nasty, brutish, and short.

—THOMAS HOBBES, *Leviathan,* 1651

NOT EVERYONE BELIEVES IN PARADISE PAST. Perfectionist theologies of nature have always inspired oppositional movements that fervently argue the contrary. Nature isn't heaven; and if not, it must be hell.

Two famous images beautifully capture these dueling perspectives. The first is a parody of the iconic "March of Progress" illustration of evolution, a series of figures that represent the development of the human species. The original illustration, from 1965, begins with "an early proto-ape" and ends with a tall, muscular, clean-shaven man. The parody—which exists in several variants—adds another human in apparent decline, hunched over a smart-phone or clutching a soft drink, shorter than his predecessor, muscles invari-ably buried under a thick layer of flab.

The other image is Alex Gregory's classic cartoon for the *New Yorker*, a simple drawing of two cavemen sitting cross-legged on the floor. "Something's not right," remarks the one on the left. "Our air is clean, our water is pure, we all get plenty of exercise, everything we eat is organic and free-range, and yet nobody lives past thirty."

Each perspective has prominent academic champions. On the side of the close-to-nature lifestyle of early humans you have the Yale political scientist James Scott, whose 2017 *Against the Grain* lays out a rigorous case for the advantages of living in "small, mobile, dispersed, relatively egalitarian, hunting-and-gathering bands."[1] According to nearly every metric Scott can think of, modern humans fare poorly. While hunter-gatherers live freely according to "natural rhythms," we moderns are slaves to time clocks and taskmasters. Hunter-gatherers have an encyclopedic knowledge of nature; most of us can barely identify a plant.[2] And, contrary to popular belief, the "uncivilized" inhabitants of the New World enjoyed "generally vibrant good health," until it was spoiled by diseases of civilization unheard of until the advent of urbanization and agriculture: cholera, smallpox, mumps, measles, influenza, chicken pox, and malaria.[3] In keeping with the "March of Progress" parody, Scott even points out that "hunter-gatherers were several inches taller on average" than early farmers, whose growth was stunted by their monotonous grain-based diets, a woeful substitute for the "varied and abundant diet" of their nomadic forebears.[4]

In stark contrast to Scott are people like Harvard's Steven Pinker, who is decidedly optimistic about progress. Pinker mobilizes reams of statistics to make his case, from record-high average life spans to record-low rates of slavery and violence. To these he adds the remarkable intellectual and cultural achievements of recent history, among them the scientific method and the spread of democracy. For him, the relative superiority of twenty-first-century life is abundantly obvious. Denying it amounts to participation in a self-loathing "morality play," wherein technophobes turn modern humans into cartoon villains for the sake of romanticizing an "ascetic harmony with nature" that never really existed.[5] Pinker sums up his Whiggish stance on our species' history with a 2016 quote from Barack Obama: "If you had to choose a moment in history to be born, and you did not know ahead of time who you would be—you didn't know whether you were going to be born into a wealthy family or a poor family, what country you'd be born in, whether you were going to be a man or a woman—if you had to choose blindly what moment you'd want to be born, you'd choose now."[6]

This general debate about the state of nature has a variety of more specific analogues. One of the most contentious concerns the connection, if any, between living naturally and treating the environment with respect. Were Native Americans "the first environmentalists," ecological sages who, in the classic example that I was taught in primary school, always used every part of the buffalo? If so, it makes good sense to seek solutions for ecological crises in their beliefs and practices. But what if this is a myth, largely invented by starry-eyed Westerners? That's the argument of anthropologist Shepard Krech in *The Ecological Indian: Myth and History*.[7] Krech marshals numerous examples of beliefs and practices from various Native American peoples that do not align with typical visions of caring for nature, including an entire chapter on buffalo hunting that discusses the practice of slaughtering herds and leaving the carcasses to rot, removing only the tongues and humps. As Kimberly TallBear, scholar of indigenous peoples at the University of Alberta, argues in her review of the book, Krech does not blame Native Americans for "failing" to live up to ecological and moral standards that are a product of modern conservationist value systems. Rather, he suggests that it's irresponsible to depict Native American values and practices as a mirror image of those embraced by members of the current conservation movement. If Krech is right, we should recognize that living in a state of nature is not necessarily synonymous with caring for nature according to a contemporary understanding of what that means, and consider the possibility that our environmental crises are better solved with technology than nostalgia.[8]

Or take war and slavery: Are they actually diseases of civilization, relatively unknown to the egalitarian bands who lived—and still live—in a natural condition of peace and harmony? Are we the monsters, and they the "harmless people"—the title of a 1956 book by Elizabeth Thomas about the Bushmen of the Kalahari Desert? If war and violence among indigenous peoples is the product of contact with the West, we should follow anthropologist Douglas Fry and look to their cultures for conflict resolution strategies. (At the time of this writing, Fry's Peaceful Societies website at the University of Alabama lists twenty-five societies where "significant scholarly literature [supports] the claims of peacefulness." With the possible exception of the Amish, none qualifies as modern or Western.)[9]

Against this view is the work of scholars who find war, slavery, and other forms of violence such as rape and torture to be ubiquitous practices throughout human history, whether in a state of nature or removed from it, pre-contact and post-contact alike. In his sweeping study *War before Civilization: The Myth of the Peaceful Savage*, archaeologist Lawrence Keeley attacks what he sees as the dominant "neo-Rousseauian concept of *prehistoric peace*."[10] He's frustrated by those who he believes ignore evidence of prehistoric war for the sake of proclaiming "the divinity of the primitive." Another archaeologist, Catherine Cameron, has documented extensively the existence of captivity in nearly every culture and historical period.[11] "Taking captives comes naturally to us," she told me. "Captives, slavery—they are as natural as the nuclear family. I think the most important social development that has happened in the last ten to twenty thousand years is the end of slavery as a culturally acceptable practice." (In the most thorough and even-handed account I have read of this debate, the Stanford neuroendocrinologist Robert Sapolsky concludes that humans are indeed making ethical progress, but not without multiple caveats, including the fact that large-scale war was virtually unknown among nomadic hunter-gatherers.)[12]

Even the question of lifespan is more complicated than it initially appears. The oft-cited thirty-year statistic, captured in the caveman cartoon, misleads by confusing *average life span* with *typical life span* (a confusion Pinker addresses, to his credit). Most preagricultural peoples did not soldier on until thirty and then succumb to disease, injury, or infection. Rather, enormous numbers of infants and children died from intestinal parasites, dehydration, malnutrition, and other hazards of life in the wild, lowering the average life span significantly. If you made it past the age of five, however, it's likely that you would have lived a reasonably healthy life well into your sixties or early seventies.

Perhaps most difficult of all is the relationship between living naturally and personal happiness. "Are Hunter-Gatherers the Happiest Humans to Inhabit the Earth?" asks the headline of a 2017 article on the National Public Radio website. The article features an interview with anthropologist James Suzman, who describes twenty-five years of living with the Bushmen of the Kalahari in his book *Affluence without Abundance*. Though Suzman is careful not to paint their lives as flawless—"it wasn't a Garden of Eden"—he nevertheless portrays the Bushmen as effortlessly participating in an enlightened state that most Westerners strive for and fail to reach: "Today people [in Western societies] go to mindfulness classes, yoga classes and clubs dancing, just so for a moment they can live in the present," he tells the interviewer. "The Bushmen live that way all the time!"[13]

Suzman's perspective is not new. In 1966, the Chicago anthropologist Marshall Sahlins presented a massively influential paper called "Notes on the Original Affluent Society," in which he described the "Zen road to affluence" of hunter-gatherers, who need very little and therefore enjoy leisure and satisfaction. Sahlins based his assertion on work by another anthropologist, Richard B. Lee, who studied !Kung San tribespeople in southern Africa. According to Lee, the tribespeople worked an average of only 2.5 days per week—a golden age of leisure compared with us workaholic moderns.[14]

But the historian Rachel Laudan cautions against romanticizing the daily lives of hunter-gatherers, which may depend on making certain forms of labor invisible. Lee's calculations fail to account for the work that goes into processing and preparing food once it has been hunted and gathered.[15] Laudan lays out everything he ignores: fifteen to twenty-two hours per week spent on butchery, cooking, and fuel collection; four to seven hours per week spent making and repairing tools; eight hours per week spent cracking mongongo nuts. (Mongongo nuts, apparently, are a serious pain to crack, not to mention having to clean off the elephant dung out of which they are sometimes gathered.)[16]

To make matters worse, scholarship on these issues necessarily requires a hefty dose of speculation. Evidence about the lives of ancient pre-agricultural humans is restricted to archaeological findings, which rarely include soft tissue and often lack the helpful explanatory context of artifacts, architecture, and written records. Consequently, interpretations are hotly contested. Did this person die from malnutrition or disease? A blow to the head or a falling rock? Population-wide generalizations are more difficult still, and the academic literature remains undecided on basic questions about life span, disease epidemiology, and population growth rates, much less questions about typical quality of life.

These issues are only somewhat easier to resolve when addressing living hunter-gatherers. Data gathering remains challenging, which leads to uncertainty about key metrics such as life span and mortality rates. Attempting to measure subjective qualities such as happiness compounds the uncertainty.

In a reanalysis of data from several happiness studies, the economists Timothy Bond and Kevin Lang discovered that conclusions about which country is "happiest" or the effects of marriage on happiness depend on how you crunch the numbers. Tweak your statistical approach and Nigeria goes from one of the happiest five countries to the *least* happy country.[17]

Not everyone is so pessimistic about the challenges of measuring happiness. The founder of the online publication *Our World in Data*, Max Roser, an economist and an expert on global trends in living conditions, argues that happiness surveys measure well-being in populations with "reasonable accuracy."[18] But even assuming he's right, there are questions about the generalizability of happiness and the wisdom of focusing on it as the most important feature of a life well-lived. While the meaning of happiness may be intuitively obvious to readers of this book, hunter-gatherer cultures often lack an equivalent term. "Bushmen have words for their current feelings, like joy or sadness," explains Suzman in his NPR interview. "But not this word for this idea of 'being happy' long term, like if I do something, then I'll be 'happy' with my life long term."[19]

More importantly, as discussed in the previous chapter, hunter-gatherers today are not "living fossils," so they only represent a tentative version of natural living, if natural living is taken to mean isolation from the physical and cultural products of industrial modernity. (As if to remind us of this, the NPR interview with Suzman is accompanied by a photograph of a young Khoisan boy dressed in plastic flip-flops and a button-down collared shirt, lounging in a metal chair.)

Attempting to control for bias does not help settle matters, since both sides can raise legitimate concerns. On one hand, the narrative of "human progress," widely accepted by scholars until well into the twentieth century, is notoriously associated with racism and cultural chauvinism. We are living in the wake of mass genocide carried out by colonial imperialists. It has been less than a century since Hitler's camps. Modern technology led to the development of nuclear weapons and human-caused climate change. The latest triumphant debunkings of indigenous and prehistoric people's positive traits can trivialize this history.

But the urge to correct for past errors can also lead to overcorrection, transforming what should be objective scholarship into a kind of penance. The sins of one culture do not entail the innocence of those they sinned against. In addition, some scholarship—like that of Cameron and Laudan—

suggests that rosy pictures of the past, instead of correcting for bias, could actually follow a long tradition of ignoring the suffering of women. After all, it is women, according to Cameron, who are much more likely to be kidnapped, enslaved, and raped after a war raid, and it is women who do most of the culinary labor that Laudan is at pains to bring to our attention. Did the !Kung enjoy days of leisurely Zen-like contentment, as Sahlins and Lee would have it? Not if you trust the first-person testimony of Bau, a !Kung woman interviewed by the anthropologist Patricia Draper. Bau favorably compares settled life with her nomadic childhood: "In the old days we ate water root for two weeks at a time. We children even then cried and said, 'Mother, Father, give me real water! I am thirsty!' That was a hard life. I lived it. I lived in the bush and I saw it."[20] Bau's perspective is not exceptional. According to Draper, "all but one of the informants stated that the transition to settled life brought about positive changes in their lives. Informants mention the arduousness of frequent moves, always being on the lookout for a new place to hunt or gather."[21] Then again, it's entirely possible that life in the bush only became hard recently, that long ago Bau would have had a much better existence, and it is the fault of settlers that nomadic life became so arduous.

The deep ideological divides around these questions have left many academics tense and confrontational.[22] Entire subfields are routinely dismissed as "reductionist" (sociobiology) or "anti-science" (cultural anthropology). Scandals and serious accusations are commonplace. Intellectual opponents are not just wrong—they are guilty of sexism, racism, reverse-racism, McCarthyism, colonialism, censorship, or "war-ifying the past." One anthropologist who works on early violence refused to be interviewed for fear I would use the material to reinforce dangerous ideas like biological determinism and the inherent superiority of modern culture. In an email, another anthropologist accused the discipline of being dominated by a "noble savage industry," in which "reputable academics like Marshall Sahlins pander to fantasies about the goodness of nature [and] small-scale, non-literate societies are treated as a species of 'magnificent wildlife' to be preserved, cherished and imitated."[23]

The urgency of debates over life in a "state of nature" may seem like a unique product of our cultural moment. Here we are, surrounded by plentiful calories and extraordinary technology, yet problems still abound. Major depression is on the rise; the natural world is in decline. Wealth inequality grows at a rate matched only by that of our waistlines. On top of that, there's our inescapable recognition that modernity has been built on the mass graves of exploited peoples, and continues to be powered by an unquenchable thirst for energy and materials secured primarily through wanton destruction of the natural world. Some, like Pinker, want to liberate us from this hell by pointing out that it is paradise compared to what came before—that our unhappiness is the product of belief in a mythic past. Others see the myth of progress as the real danger, to be corrected only by embracing lifeways uncorrupted by modernity.

But surprisingly, none of this is uniquely modern. Despite variation on the details, the general form of this argument is remarkably cross-cultural and transhistorical. Characterizations of the state of nature as paradise or hell have a long and distinguished history. Mythologizing life in the past is one of the most common ways people make sense of the present, whenever that present might be. And so, to understand why we invest depictions of hunter-gatherers and Paleolithic man with such significance, we must first examine why others before us did the same.

In the epigraphs to this chapter, John Dryden and Thomas Hobbes stake out positions that are easily recognizable today: the "noble savage" who lives free from civilization's fetters; the state of nature that is "solitary, poore, nasty, brutish, and short." Clearly the dichotomy is not a product of modern discontents and biases, since the two men were writing nearly four centuries before World War II, nuclear weapons, and the recognition of human activity as an existential environmental threat.[24]

Nor is it uniquely Western. Go back another two millennia, and you find these two quotes from Chinese classics that repeat the same dichotomy:

> The men of old dwelt in the midst of crudity and chaos; side by side with the rest of the world, they attained simplicity and silence there. At that time the yin and yang were harmonious and still, ghosts and spirits worked no mischief, the four seasons kept to their proper order, the ten thousand things knew no injury, and living creatures were free from premature death. Although men had knowledge, they did not use it. This was called the perfect unity. At this time, no one made a move to do anything, instead being constantly self-so [*ziran*, usually translated as "natural"].
>
> —*Zhuangzi*, c. 300 BCE[25]

In ancient times, people ate vegetation and drank from streams; they picked fruit from trees and ate the flesh of shellfish and insects. In those times there was much illness and suffering, as well as injury from poisons.

—*Huainanzi*, c. 200 BCE[26]

These archetypal visions of life in a state of nature—one of which exaggerates its virtues, the other its vices—are born of two conflicting explanatory tendencies. The first is the desire to explain suffering with reference to a previous model of perfection. Myths of Eden and the Fall are not solely the province of Abrahamic religion. Variants exist in Greek, Chinese, Egyptian, and Indian traditions. During the mythic Hindu epoch known as the Satya Yuga there was no need for agriculture, no one fell ill, and the weather was always nice—until humans developed disordered desires. Even the pre-agricultural peoples who star in modern golden age myths had their own versions of the same. A myth from the South American Bororo people describes "days when diseases were still unknown and human beings were unacquainted with suffering."[27]

By making sense of our suffering, this mythic framework alleviates it. In the words of religious studies scholar Mircea Eliade, "suffering is perturbing only insofar as its cause remains undiscovered."[28] Though Eliade overstates the case—certainly suffering remains perturbing after being explained—he is right to observe that cosmic explanations make worldly woes easier to endure. In addition to creating meaning, such explanations also offer a general prescription for redemption. If our current state, whenever it might be, represents a fall from a natural state of perfection, the solution to suffering lies in recreating that state to the best of our abilities. Are children disobedient? Then we should raise them as we did at the dawn of time. Are we sick? Then we should eat and drink what the first humans ate and drank. Are we plagued by war and violence? Then we should organize ourselves as we were organized in the ancient days—or, following the implications of Dryden and the *Zhuangzi*, we should reject artificial organization altogether, refuse to obey "base Laws of Servitude," and live freely in harmony with the four seasons.

Opposed to belief in a paradisiacal state of nature is the desire to see one's current culture as the pinnacle of human progress. We observe how technology improves over time and how scientific knowledge builds on past errors. Once, the earth was flat and dragons lay at the edges of maps. Now we photograph the truth from outer space. Extrapolated backward, this means increased proximity to the state of nature is directly correlated to ignorance. Moreover, as Martin Luther King Jr. (quoting Theodore Parker) reminds us, we like to believe that "the arc of the moral universe is long, but it bends towards justice." Despite short-term setbacks, most people see the developments of democracy, gender equality, the abolition of slavery, religious tolerance, and universal human rights as progress, marked improvements over the moral codes of the past that did not include them. According to this myth we began not as innocents but animals: ignorant, amoral, instinctual creatures who lived in a natural state of competition.

In addition to making people feel good about their culture, the Hobbesian perspective justifies forcing it on others. Mythic dehumanization of those in a state of nature reframes government control as a blessing and colonization as a gift. In *The Myth of the Noble Savage*, Ter Ellingson catalogues the terms used by colonialist authors to describe indigenous cultures across the globe and justify their subjugation. These were the "lowest," "most savage," "most brutal," "wildest," "least human," "most degraded," "most degenerate" specimens of humanity, and therefore could be treated like children, at best, and wild beasts at worst.[29] Writing in 1632, one missionary introduces his book about Native Americans with the following description: "Savagery and brutishness have taken such hold that the rest of this narrative will arouse in your souls pity for the wretchedness and blindness of these poor tribes. . . . You shall see as in a perspective picture richly engraved, the wretchedness of human nature, tainted at the source, deprived of the training of faith, destitute of morality."[30]

Such characterizations underwrite the ideal of the White Man's Burden, as Rudyard Kipling infamously titled his 1899 paean to imperialism. Originally intended to justify the American annexation of the Philippines, his poem captures neatly how exploitation of native populations depended on seeing them as "new-caught, sullen peoples, / Half-devil and half-child," living miserable lives of "famine" and "sickness" that only civilization could redeem.[31]

Neither of these myths matches up with reality, which results in severe cognitive dissonance for people who believe in them. The very same missionary who described Native Americans as wretched and immoral grudgingly admitted that "they reciprocate hospitality and give such assistance to one another that the necessities of all are provided for without there being any indigent beggar in their town and villages; and they considered it a very bad thing when they heard it said that there were in France a great many of these beggars, and thought that this was for lack of charity in us, and blamed us for it severely."[32] Most every European visitor to the so-called New World noted the vibrant healthfulness of the natives, including Christopher Columbus, who described the inhabitants of the Canary Islands as "all without exception [having] very straight limbs, and no bellies, and very well-formed."[33]

In the same manner, those who want to redeem the state of nature invariably bump up against its shortcomings. James Scott, while arguing for the relative abundance of food and leisure in hunter-gatherer contexts, also suggests that they lived lives of scarcity and hardship. "Slavery," he admits, "was common among manpower-hungry Native Americans."[34] To account for the lower reproductive rates of hunter-gatherers, Scott cites a "combination of delayed weaning, abortifacients, and neglect or infanticide." Bracketing the morality of infanticide (which Scott does not address), the practice points to an existence so precarious that an extra child could prove ruinous.[35]

Still worse, the comparison of one's own culture with a single "state of nature" requires oversimplification and generalization that renders any conclusions highly suspect. Hunter-gatherers and indigenous peoples are not monolithic entities. Their circumstances and lifeways are incredibly diverse, and assuming they are basically identical can reinforce the homogenizing essentialism of words like "savages" and "barbarians," even if modern terms seem gentler. Similarly, speaking of modern Americans as an undifferentiated population ignores how average life span can differ by as much as twenty years depending on one's zip code. The result is cherry-picking of examples and metrics—should we compare rates of infant mortality or chronic illness?—that end up supporting the myth you already believe.

Our desire to take stock of the present by comparing it with the past makes these comparisons irresistible. We seem compelled to continue the ancient tradition of asking, "Did our hunter-gatherer ancestors have it better?" (a question taken verbatim from a recent headline in the *New Yorker*).[36] Even after completing much of the research for this book, I still felt the compulsion. Ridding myself of it completely required a visit to an indigenous community in Peru, where I finally realized the impossibility of answering this question, and the potential pitfalls of asking it.

I found out about the Matsigenka during an online search for parenting advice. These searches became a part of my life from the moment of my daughter's birth in 2012. *Is the "cry-it-out" method okay for sleep training? What's the best way to discipline a toddler? When should we start potty training?* Invariably, the web led me to articles about natural parenting—that is, parenting in a state of nature. "Best Practices for Raising Kids? Look to Hunter-Gatherers," counseled the anthropologist Jared Diamond in a 2012 book excerpt published by *Newsweek*, which I discovered one sleepless night after being awakened by my six-month-old.[37] Turns out that hunter-gatherer children sleep near their parents for years (unlike my daughter, who slept in a crib that wasn't even in our bedroom). In a state of nature, cry-it-out would be tantamount to child abuse—apparently hunter-gatherers can't bear to let their children cry for even an instant. As for potty training? No diapers, obviously: you simply spend all day with your children and respond to their cues, an approach that has been dubbed "elimination communication" and advocated by celebrities including Alicia Silverstone, who used it successfully with her son, Bear. "It's the most natural, primal thing," she told *People* magazine.[38]

When our daughter Hazel was about to turn five, my wife and I started wondering about household chores. How much should be required of children, and what kind of approach should be taken when introducing responsibilities? Despite being sweet and generally well-behaved, Hazel would resist cleaning her room and folding her laundry. I'm a demanding parent, I thought. Maybe it was too much to ask?

Trawling the internet for advice, I stumbled across an article titled "Spoiled Rotten," in which Elizabeth Kolbert reported on a study of the differences between middle-class Los Angeles children and their counterparts in a Peruvian tribe.[39] I read it. I read it again. I called my wife over and read it to her. And then we reflected seriously on the virtues of living like a hunter-gatherer, and the harm we might have done with diapers, cry-it-out in the crib, and our other unnatural approaches to child-rearing.

Unsurprisingly, the LA children were like Hazel. "No child routinely performed household chores." Check. "Often, the kids had to be begged to attempt the simplest tasks; often, they still refused." Check. But the Matsigenka children? In the study, anthropologist Carolina Izquierdo described typical behavior that was nothing short of miraculous. Unasked, a girl named Yanira would sweep sand off sleeping mats, help stack leaves, fish for crustaceans—and then clean, boil, and serve them herself. Yanira was six. Other details were no less astonishing, like how "toddlers routinely heat their own food over an open fire," and "three-year-olds frequently practice cutting wood and grass with machetes and knives." It sure seemed like there was something to parenting in a state of nature. And what if it wasn't just parenting? What if life in general was blissful? I had to see for myself.

Immediately I contacted Izquierdo to ask about the advisability of meeting members of a Matsigenka community in person. My primary concern was with the ethics of "cultural tourism." There is an industry built around tourists gawking at the quaint practices of isolated indigenous cultures, and it bothered me that I might be a participant. Unlike anthropologists who live for years with the people they study, I planned to spend only a few days in the jungle. I do not speak Matsigenka, so I would be forced to communicate in Spanish. But Izquierdo assured me there wasn't anything intrinsically wrong with visiting another culture to learn about them, provided that your goal is to encounter its members as fellow humans, rather than exoticize them as objects of curiosity. And so I proceeded to seek out a guide with Matsigenka contacts, booked my plane tickets, and began reading up on the history of a people my wife and I knew only as the ideal parents we wanted to be, but couldn't.

The Matsigenka community of Huacaria lies nestled in the Amazon rainforest, eight bumpy hours from the mountain city of Cusco, where tourists usually spend a night before heading to Machu Picchu. When my wife and I arrived, Scott, our guide, warned us about the Matsigenka we were going to meet. Even within a specific tribe there are many different ways of life. Since we lacked a permit, there was no way to meet those Matsigenka who choose to live deep in the jungle and experience virtually no contact with outsiders. In Huacaria, he said, most people wear jeans and T-shirts. "Foreigners want to see traditional clothing, and sometimes indigenous people dress up for them, even though that's not what they usually wear," he told me. "We are going to meet real Matsigenka, but I don't want you to be disappointed if they don't look the way you think they should."

Real Matsigenka. Without intending to, Scott had tapped into another concern of mine, about the state of nature's citizenship requirements. While researching the Matsigenka, I noticed that in most photos they were dressed as Scott described, in machine-made clothing, not traditional *cushmas* woven from nearly three miles of laboriously handspun cotton. (In another version of golden age mythology, the Matsigenka tell of a time when women did not have to suffer to obtain cotton thread, instead getting it from Heto, the "spider woman" who spun it effortlessly from her bellybutton.)[40] The Matsigenka I was to meet dressed like the Khoisan boy photographed by Suzman and featured in the NPR article. Were they still living naturally, then? I'd asked myself the same question about Maasai herders who have opted to outfit their traditional huts with solar panels, and now use cell phones to download weather forecasts and flirt on WhatsApp. To talk about a culture as "traditional" or "hunter-gatherer" or "natural," it seems like there needs to be a set of essential qualifying features, and disqualifiers. Specific practices themselves need to be parsed. Is hand-spinning cotton part of living in a state of nature, or is it a technological step toward renouncing one's citizenship?

In a certain sense, no one lives in a state of nature, not anymore. Even for those rare tribes that have little to no contact with outsiders, circumstances are vastly different than they once were. Most employ at least some form of borrowed technology like metal axes, machetes, plastic containers, and aluminum cookware. The territories where they live have been radically circumscribed by the political and economic activity of nation-states. Pollution, modern hunting practices, and climate change all mean that the ecosystems they live in are generally quite different from what they once were. It seems entirely inappropriate to call any of this a state of nature, altered as it is by the activities of post-agricultural, industrial human civilization.

Upon arriving in Huacaria, however, it was undeniably obvious to me that the Matsigenka are closer than I to living in a state of nature, wherever its blurry borders happen to lie. Yes, they have solar panels, recently commissioned by Don Alberto, Huacaria's shaman and a close friend of our guide. Yes, they use plastic containers to wash manufactured dishes with chemical soaps in newly installed communal sinks. And yes, there is a one-room schoolhouse where non-Matsigenka teachers instruct children in Spanish. But many basics of everyday life emerge directly from natural systems—plants, animals, and rhythms created and maintained by primal, non-human forces. I dined in a typical home crafted entirely from native jungle materials: a palm-thatched roof and walls of bamboo slats lashed together with strangler fig vines, a centuries-old technique. Bows are made of *chonta*, a hard palm, and used to hunt monkeys, peccary, and other forms of game. Just outside Huacaria, Scott, himself an accomplished botanist and herbalist, smacked his machete blade into a tree heavily scarred with previous blows. "*Sangre de grado*, instead of Band-Aids," he explained, rubbing the blood-red sap in the palm of his hand. Rich with natural latex, the sap became white foam, and Scott applied it to a small cut on his leg, where it dried to form an impermeable barrier.

When I juxtaposed Matsigenka homes with my air-conditioned house; their food with supermarket produce flown in from far corners of the earth; the *sangre de grado* with Band-Aids made according to a process I do not understand from materials whose origins are shrouded in mystery—when I looked at my life and Matsigenka life side by side, the difference between natural and unnatural living, a spectrum though it might be, was as clear as the difference between a *chonta* growing in the jungle and a plastic plant in an office lobby.

Less clear, though, was the relationship between living closer to the natural end of the spectrum and living a good life. All around me was evidence that the people of Huacaria were actively choosing to integrate less natural practices. In halting Spanish, an elderly Matsigenka woman told me how grateful she was for running water. "As a little girl my stomach always hurt," she said. "And we had to wash clothes in the river. Now it is easier." Another man looked at me with confusion when I asked if he was happy about having electricity. "Yes," he said flatly, as if explaining something simple to a child. "Now we can see at night." There was the school, where outsiders were invited to teach children the technologies of reading and writing. And, in addition to solar panels, Don Alberto had also secured government funding for the construction of an artificial fish-farming pond and tourist lodging. He hoped that tourists would bring in money, which could be used to purchase everything from salt to automobiles. The Matsigenka who lived so much closer to the state of nature than I—they were choosing to distance themselves from it.

Yet there was also evidence of how natural living can be advantageous, some of it confirming the observations of early colonial sources. The poorer parts of Peru's cities are strewn with litter and many residents suffer from homelessness; in Huacaria, homelessness is unknown and the only litter is imported. Every person I saw in Huacaria looked physically fit, or as Columbus put it, "very well-formed." Less than a half hour away from their town, the forces of civilization are destroying the environment on a scale that's unimaginable to the Matsigenka, whose waste products biodegrade instead of persisting for thousands of years. Were they making a mistake, falling for the illusory benefits of modernity only to end up trapped in a worse life than before, the latest participants in the world's perdition?

A long conversation with Don Alberto, arguably the architect of Huacaria's modernization, did little to resolve these tensions. A shaman by training, Don Alberto had spent fifteen years in the city before returning home. In fluent Spanish he mourned the slow erosion of natural living in all its forms: his knowledge of local herbs and plants that might not survive past the next generation; the devastation of the jungle and its inhabitants for the sake of farms and animal agriculture. He told me the new fish pond was necessary because commercial overfishing had made it impossible for Matsigenka to continue their old fishing traditions.

"New inventions have done much harm," he said. "Civilized man is sick. He looks up and spins his fingers to change the world. Then he comes to us and asks for help with his problems, his new diseases. He is the enemy of Pachamama [Mother Earth]."

"Is technology bad?" I asked, thinking about all the new inventions he had brought to Huacaria, the electricity and the running water.

"Yes!" Don Alberto answered. And then, a second later, as if sensing my follow-up question, "Well, no. Light allows me to read and write. Cell phones can connect us. Computers can create things. Yes and no."

There, in the Peruvian Amazon, a Matsigenka shaman with a cell phone on his belt came to the same paradoxical conclusion reached by so many other cultures. Is life worse in a state of nature? Or did our hunter-gatherer ancestors have it better? *Yes and no.*

On our way home from Peru my wife and I had a layover in Florida. Airports are not the most pleasant places, but we still felt grateful for the boons of civilization: the bounty of food, the absence of stinging insects, the ability to learn about the world via the internet and the television at our terminal. And so it was, watching TV, that we found out the celebrity chef and author Anthony Bourdain—a hero to both of us—had committed suicide by hanging. Stunned, we then learned that just two days earlier, the fashion designer Kate Spade had done the same. Coverage of their tragic deaths made frequent mention of rising rates in depression and suicide. The shaman's paradox had followed us home.

It's easy to find research that blames these rising rates on modernization. One scholarly review identifies depression as yet another disease of civilization, right along with James Scott's list of viral infections. Hunter-gatherers are happy, until they leave the state of nature. "The Ik of Uganda purport-edly become more depressed upon shifting from hunter-gather to agricul-tural practices," the review reports. "After indigenous circumpolar peoples rapidly modernized, there was a rampant incidence of diabetes and suicide rates tripled within a decade."[41] Multiplied across cultures, the evidence is damning: "A cross-cultural analysis of community women in rural Nigeria, urban Nigeria, rural Canada, and urban US found the degree of moderniza-tion to correlate with a higher prevalence of depression in a dose-dependent manner."[42] The analogy is clear: negative effects of modernization are dose-dependent, just like a poison. By distancing ourselves from natural ways of life we are slowly poisoning our souls.

In the past, the same argument was made by citing reports of Europe-ans defecting to live with indigenous people. "If these savages are as un-happy as it is claimed," Rousseau asked pointedly, "by what inconceivable depravity of judgment do they refuse steadfastly to civilize themselves in imitation of us and to live happily among us, whereas one reads in a thou-sand places that Frenchmen and other Europeans have voluntarily found refuge among these peoples?"[43] As one French nobleman wrote in 1703, "I envy the state of a poor Savage. . . . I wish I could spend the rest of my life in his Hutt."[44]

But as Ellingson observes in *The Myth of the Noble Savage*, the French nobleman who wrote those words did not actually spend the rest of his life in a hut, and Rousseau did not renounce civilization to live in a state of nature. In fact, Rousseau vehemently rejected that possibility. He readily admitted that he was unwilling to "nourish himself on herbs and nuts," and mocked those who would give up city living for the forests. For Rousseau, the state of nature was most useful as a thought experiment, a construct that "perhaps never existed."[45] Did it help him understand the flaws of his own culture more clearly? Undoubtedly. Should we strive to return to it? Just asking that question would have struck him as misguided, since it makes no sense to return to a mythic time.

This is the pitfall of attempting to compare modernity with a state of nature, or my own culture with that of the Matsigenka: it encourages us to confuse mythic constructs with reality. As Rousseau was well aware, the most obvious of these mythic constructs is the state of nature itself. Less obvious is the mythic objectivity required for making comparisons between radically different ways of life, as if there were a cosmic balance onto which we can place all the shortcomings of two cultures and then stand back to see which one rises.

Let's assume depression and suicide were in fact unknown to hunter-gatherers, and only emerged with modernity. Place that weight on the side of modernity. Add to it the weight of every disease of civilization. Add, again, the weight of the environmental devastation wrought by advanced technology, and the weight of large-scale war. But on the other side, place the bodies of infants and children who died more frequently than they do today. Alongside them lay the mothers who died in childbirth, and the elderly who miss out on another decade. According to Nicholas Blurton-Jones, an expert on Hadza hunter-gatherers, "None of us [anthropologists] believes we have seen any 90+ years old hunter-gatherers, and there are many fewer 80-year-olds than here in southern California."[46]

Keep adding until you can't think of any more metrics. Now step back and see what happens. For me, the existence of ninety-year-olds is hard to overvalue, since my own father is ninety-one. Others will value it less. And therein lies the problem: while science can tell us objective truths about suicide rates and infant mortality rates and deforestation rates, it cannot provide a balance on which to weigh them objectively.

Moreover, cultures—like individual humans—aren't the simple sum of easily measurable metrics. Any worthwhile balance will need the capacity to weigh cultural sins such as genocide and slavery, as well as cultural knowledge and artistic production. But according to what standards? How much does accepting the practice of infanticide lower one culture's standing? Does it weigh more or less than accepting the practice of slavery? What about ignorance of local plant and animal species, for which Scott chastises modern humans? Is that worse than ignorance about the germ theory of disease, or the shape of the planet?

Along with the mythic state of nature and the mythic balance for comparing it with modernity, there are also mythic utopias that follow from the comparison. Once we decide hunter-gatherers "had it better," all that remains is to integrate their practices into our lives as much as possible. Live off the grid, go barefoot, sleep with your kids until they're four. Presto: the dose-dependent poison of modernity is reduced, and happiness goes up! Conversely, if we decide that Hobbes was right, it follows that everyone, everywhere, should be modernized as quickly as possible, the better to distance them from a hellish state of nature. It's *Lord of the Flies* made real, in

which the children—"brown, with the distended bellies of small savages"—
are saved by the deus ex machina of civilization, a uniformed naval officer
with a revolver at his side. Use education to wipe out primitive superstitions,
hand out laptops and cell phones, and watch the miserable savages flourish!

All forms of utopianism blind us to complexity. Even if it were true that
modernization makes people happier, the question of how to go about it is
immensely complicated. Should the Peruvian government accelerate Huacaria's industrialization, encouraging the Matsigenka to trade their bows for
guns, their oral creation stories for science books, their handmade homes for
cinder-block apartments? What kind of agency should the people of Huacaria have when it comes to decisions about the direction of their culture?
Utopianism about progress tends to gloss over the long and sordid history
of failing to ask these questions, and the calamitous results of those failures.
It usually privileges the powerful—who, in the end, control the definition
of progress—and rarely solicits the input of those whose cultures stand to
lose the most.

The great Peruvian author Mario Vargas Llosa captures the very worst
version of this attitude in his novel *The Storyteller*, which follows a young
man named Saúl who leaves society to live with the Matsigenka. The narrator, a classmate and a friend, becomes frustrated with Saúl's intense affection
for Matsigenka culture, and challenges him on the value of preserving their
way of life. Yes, there's something intriguing about indigenous customs, he
admits, but progress is obviously more valuable. From there it's one step
to the queasy logic of cultural genocide: "If the price to be paid for development and industrialization for the sixteen million Peruvians meant that
those few thousand naked Indians would have to cut their hair, wash off their
tattoos, and become mestizos—or, to use the ethnologists' most detested
word, become acculturated—well, there was no way around it."[47]

However, romanticizing hunter-gatherers is no less problematic. When
the state of nature represents purity, then every fruit of modernity becomes
a potential source of pollution. Food is toxic. Cribs are abusive. Unnatural
lifestyles destroy our bodies. Technology destroys our souls. This mind-set
conduces to deep existential anxiety, which leads to the desperate piecemeal
adoption of seemingly natural practices. In certain cases the piecemeal approach can work, but it fails to consider how the efficacy of practices depends on the culture in which they are embedded. Going barefoot is a very
different experience when you're walking on hot concrete or broken glass.

Child-rearing "like a hunter-gatherer" requires a network of cultural re-sources that are unavailable to many of us non-hunter-gatherers. Research strongly suggests that an indispensable feature of child-rearing among hunter-gatherers is the presence of kin who can help care for the children (the "grandmother hypothesis"). Their presence can only be guaranteed within a specific social structure supported by a specific set of values. "Isolat-ing the nuclear family, as we do, can be very stressful," said Kristen Hawkes, one of the anthropologists responsible for developing the grandmother hy-pothesis, in an interview with me. "But the remedy isn't obvious. Our drive for separate nuclear families is part of the current economy. Changing it isn't something that any individual can easily do."

Elimination communication might work for Alicia Silverstone, who al-most certainly has domestic help—but will it really be the best approach for working parents without family nearby? What about society? Should we embark on a large-scale reorganization that moves us toward geographically proximate kin networks?

Archetypal thinking cannot answer these questions. States of nature, whether paradisiacal or hellish, do not have room for shamans with cell phones, or parents trawling the internet for advice about their kids' chores. Myths swallow up authentically complicated ways of life. They are easily digested substitutes for the more difficult stories of real human beings and cultures, which exist on a continuum between natural and unnatural, and cannot be evaluated according to their distance from one of those poles.

Recognizing the inadequacy of mythic binaries can help us abandon a theological version of nature. But religious stories take many forms. Some-times they are just straightforward narratives, expressed in symbols, written and oral. More often, however, we become protagonists in the stories. We live out the plots and morals, commemorating them with sacred activities— which, in religious studies, are known as rituals, and require a different criti-cal approach.

PART II

Ritual

Before getting married I had a different last name. My "bachelor" name is Dagovitz and my wife's maiden name is Levin. We liked how our culture's traditional marriage ritual uses a shared last name to symbolize the union of two people, but we did not like how the woman's previous identity symbolically disappears into the man's. Hyphenation felt like a one-generation fix that still required one of our names to take pride of place.

And so: Levinovitz.

Violation of ritual norms always comes at a cost, and our last name was no exception. First I had to pony up $400 to the state of Illinois, because while a woman can take a man's name for free, most states demand payment if a man wants a new name—an expensive reminder that legal rituals are bound up with cultural rituals. But that was nothing compared to the fallout with my parents. They aren't especially bound to tradition, so I expected them to approve of our decision or at least be indifferent. Instead they were upset, complaining about not knowing how to properly address their mail to us and enacting a kind of taboo on saying the hybrid monstrosity that lasted for weeks. No matter that according to family lore, "Dagovitz" was itself invented on Ellis Island by my grandfather. Just think: "ovitz" is a bastardization of "ovich," which means "son of" in eastern Slavic languages. By taking on my new name, it was as if I had ritually disowned my birth family and become the son of my in-laws. Had I been in their position I probably would have reacted identically.

Though some people might be more tolerant of this particular break with ritual, there is no one who doesn't actively affirm the ritual importance of names. We address authorities by their last names and use honorifics such as Judge and Doctor; we have special nicknames for our loved ones. As Rumpelstiltskin knew well, names have mystical power. Even when rituals aren't centrally concerned with naming, they often depend on the correct use of names. Invoking the right name—one's own or that of a deity—makes rituals efficacious, when signing a contract or accepting Jesus. And as values and power structures shift, so do the names to which they gave birth: Indians become Native Americans; the era after Christ becomes the "common era," not *anno domini*.

The ongoing debate over whether to shift from BC/AD to BCE/CE showcases how the ritual of using a name, in this case the name of a time period, has existential and religious significance. The alternative approach dates back to mid-nineteenth-century Jewish scholars who, understandably, wanted to avoid appearing to endorse the "our lord" that is built into *anno domini*. (Printed names are important in Judaism, and many Jews won't write out "God," instead using a dash: "G-d.") But attempts to standardize the alternative have met with staunch resistance. When, in 2011, the BBC decided to allow writers editorial discretion about which form to use, the move drew furious responses. "BBC Turns Its Back on Year of Our Lord: 2,000 Years of Christianity Jettisoned for Politically Correct 'Common Era,'" read a headline in the *Daily Mail*.[1] The article pointed out that commissioning editor Aaqil Ahmed, a Muslim, appeared to be behind the decision, and accused him of confusing readers with "alien language." According to the Southern Baptist Convention's official statement on changing calendrical nomenclature, "this practice is the result of the secularization, anti-supernaturalism, religious pluralism, and political correctness pervasive in our society."[2]

Unpacking the ritual importance of names shows how rituals crosscut multiple cultural domains: law, religion, linguistics. This inevitably leads to definitional ambiguities. As a religion professor, I'm often asked by my students what counts as a ritual. *Must a ritual be religious? Is saying "bless you" after someone sneezes a ritual, even if you're an atheist?* My answer, though perhaps unsatisfying, is that academics have yet to settle on a definition. Most traditional rituals originated in times and cultures that did not distinguish theoretically or practically between religion, philosophy, science, and medicine. Like many of his contemporaries, the seventeenth-century astronomer Johannes Kepler believed that God was a divine geometrician, and understood his own mathematical study of the stars as veneration of God's divine handiwork. Scientific inquiry *was* religious ritual. In this he resembled Chinese astronomers from the same period, who believed the rhythms of the stars were part of a universal system that involved everything from the proper time for marriage to the functions of the human body. The system's rules governed ethics along with physics, culture along with nature.

The implications of this worldview for medicine are clear: ensuring good health requires adherence to the rules of the cosmic order. It makes sense that at this time in China, as in Europe and the rest of the world, medicine was inseparable from religious beliefs about virtue and sin, and healing rituals were as likely to involve the soul as the body. (When Copernicus, Kepler's predecessor, studied medicine at the prestigious University of Bologna, required classes included "medical astrology.")

A defining aspect of modernity has been the division of these disciplines, which challenges the idea of a single, discoverable, unifying principle of order. For the vast majority of us, natural disasters and sickness are no longer a function of someone's sinfulness. A century ago it might have seemed reasonable for the Florida minister Kevin Swanson to blame hurricanes on homosexuality, or for the Indian health minister Himanta Biswa Sarma to blame cancer on "sins in a past life."[3] But because they made these claims in 2017, public outrage at both men was swift and widespread. Solving meteorological and medical problems is not a matter of ritually purifying our culture, or ourselves, and suggesting otherwise insults the victims and our intelligence. Instead of looking to pastors and Brahmins for guidance, we rely on scientists and physicians: they, not religious figures, are the ones who tell us what steps we should take to protect our cities and heal our bodies.

The shrinking sphere of organized religion's explanatory authority, along with the strong association of ritual with religion, can easily lead to the illusion that ritual itself is becoming less important in people's lives. As my family's reaction to my name change demonstrated, however, rituals need not be explicitly religious to exert power and make meaning. Although academics cannot agree on a succinct definition of ritual, there's consensus that it encompasses much more than the practices of modern organized religion. Our lives would be unimaginable without rituals: initiations, sporting events, graduations, and holidays, not to mention less consequential ones such as handshakes, honorific titles, and even holding the door for others, which my school, James Madison University, proudly identifies as one of its defining rituals. "Students often describe their experience as one where we 'hold the door open' for each other," reads the official school website. "We see this as both literal—reflecting a warm and welcoming community—and figurative, where we open opportunities for our students by fostering the cultivation of ideas in and beyond the classroom." Rituals, whether simple or complicated, once-in-a-lifetime or everyday, are lived codes that reinforce the temporal, social, and moral frameworks that define our identities.

When rituals are violated, as my wife and I violated the ritual of familial naming, it can result in the violators being perceived as immoral, unclean, dangerous, and "other." This is most obvious in the violation of taboos, which are also known as negative rituals. If someone walks around naked in public, that will provoke an extreme reaction far out of proportion to the (nonexistent?) danger posed by seeing someone naked. But by violating one of our most basic negative rituals, the public nudist is declaring that they are not bound by the public order—maybe he also picks his nose, or sleeps with his relatives, or thinks that random violence is permissible!

Just as the violation of ritual can indicate disregard for order, the performance of rituals can function to reestablish order when we feel disempowered, disconnected, and insecure. Meditation, yoga, gardening, a solo hike, a family dinner without devices—performed regularly, such activities are popular means of cultivating proper order, stripping away superficial concerns and reminding us what really matters. Though they may sound cliché, especially when their benefits are overstated, I know of no one who doesn't practice some version of an ordering ritual.

Strikingly, many of these secular rituals are bound up with nature. Hikes, gardening, birdwatching, camping, fishing, rafting—for the majority of those who practice them regularly, the activities are not mere hobbies. They serve a more important purpose, helping to correct an existential imbalance that is popularly referred to as "nature-deficit disorder." Many of us work long hours under artificial lights in built environments, earning money to spend on electronic devices that provide endless indoor entertainment to us and our children—but it's believed that this comes at the cost of our well-being, which requires contact with the natural world, a need that the biologist E. O. Wilson termed "biophilia." Along with traditional outdoor activities, detoxes are an increasingly popular way of dealing with the negative results of unnatural living, prescribing a regimen of natural foods and natural products to purify our bodies, and our souls.

But connecting with nature, like connecting with God, requires clarity about what the sacred name describes. If you pray to the wrong deity, the healing won't work; to purify the world in the name of a false idol is to desecrate it. The same is true for rituals that center on nature. This explains why debates over what counts as "natural" are so fierce, and why those debates are at once scientific and theological—scientific, because they are restricted to empirical evidence, and theological, because the right way to perform our sacred rituals depends on the outcome of the debate. When these rituals are integral parts of our lives—as forest walks are for me, having grown up cherishing my time with northern California redwoods—it can be difficult to step back and assess them impartially. Nevertheless, if we believe in the value of nature, if we believe in the need to protect it and participate in it, then we must be open to questioning the structure of our rituals and the meaning of the names that make them work, for the sake of ourselves, our communities, and the natural world.

CHAPTER 4

Hey Bear!

THE GRIZZLY APPEARED OUT OF NOWHERE, barreling full speed across one of Yellowstone's few paved roads. Keen-eyed even in the shadowy dawn, my guide, the wolf-watcher Ilona Popper, saw him just in time—a juvenile male, she told me afterward—and shouted a warning. I screeched to a halt just yards from the massive creature, which was moving faster than my car had been only moments before. (Safety literature emphasizes that grizzlies top out at 35 mph, so running is foolish in a close encounter.) As he disappeared into the low grassy hills I took some blurry photos through the window, hands shaking with exhilaration at the encounter and relief at the bear's safe passage.

During the rest of my time in the park I never again saw a grizzly, though I did join knots of roadside hopefuls chasing rumors of an earlier sighting, patiently scanning the landscape with binoculars and spotting scopes. These were diverse crowds: tourists like me alongside professional photographers, naturalists, and dedicated wolf-watchers taking a break from their usual routine. The sight of a giant bear ambling through its native homeland never fails to awe, not even the locals and rangers who have seen them several times before.

Yellowstone is neither a zoo nor a nature documentary, so animal sightings are a privilege, not a given. Before setting out into the park, Ilona had cautioned that I might not see a grizzly, or wolves for that matter, which have become a huge attraction since their successful reintroduction in 1995. Her disclaimer proved unnecessary. In addition to that first magnificent bear I saw wolves, bald eagles, and, incredibly, a newborn bison just minutes old, the glistening placenta still dangling from its mother as she gently nudged her baby back toward the herd, away from potential predators.

Yet even if I hadn't seen any of these animals, it's hard to imagine being disappointed. My primary experience was of the place as a whole. The real gift of charismatic megafauna, as the ranger and Yellowstone historian Paul Schullery has observed, is to get us out there, waiting, paying attention, so we can "soak up all the other things that are always going on."[1] Out there, you cannot fail to notice that life and death are going on. Look down and there are sure to be bones nearby, picked clean and sun-bleached, dissolving slowly into the soil that will one day nourish the bison calf with a diet of grass and sedge.

In my regular life, the beginning and end of the story are revealed only in flashes: the birth of someone's child; the death of a loved one; buds in spring; maggoty roadkill. In Yellowstone, beginnings and endings are omnipresent. The world, it reminds me, is always a nursery and a cemetery. I am there with the bison and the bones, the grasses and the sedge, and I see the end of my own story. Nature's raw honesty exceeds any frame and every binary.

As long as national parks have existed, people conceived of them in religious terms. In his book *Discovery of the Yosemite*, the nineteenth-century explorer Lafayette Bunnell described the valley as hallowed ground. Yosemite wasn't just beautiful, it was holy, "the very innermost sanctuary of all that is Divine in material creation," a place where visitors could "commune with Nature's God."[2] When his companions failed to behave respectfully, Bunnell reacted as one would in church. "It may appear *sentimental*," he wrote, "but the coarse jokes of the careless, and the indifference of the practical, sensibly jarred my more devout feelings . . . as if a sacred subject had been ruthlessly profaned, or the visible power of Deity disregarded."[3] (Despite revering the valley, Bunnell had no problems profaning the Native Americans whose presence long predated his "discovery," calling them "naturally vain, cruel, and arrogant.")[4]

Today people pay over 350 million visits to US national parks each year, visits that might be better characterized as pilgrimages. The pilgrimage is an archetypal religious ritual in which individuals seek spiritual significance and renewal through travel to a holy place. This experience is precisely what national parks have always promised, as Lynn Ross-Bryant meticulously documents in her book *Pilgrimage to the National Parks*.[5] We go to "get away from it all," to be awed and humbled, surrounded by grand systems in which we had no hand. Hikes lead to overlooks dubbed "Inspiration Point" in Yosemite, the Grand Tetons, and Yellowstone, recalling the original meaning of inspiration: to be filled by divine breath with life and spirit. "Better men and women will result from their visits to the great open breathing spaces," declared the director of the National Park Service in 1921.[6]

Pilgrimages are constituted by smaller rituals: prayers, sacrifices, ablutions. In a sacred place, pilgrims can communicate directly with divinity. It's an opportunity to demonstrate devotion through giving up time and resources. Cleansed of earthly concerns, you reconnect with what matters. In Yellowstone, devoted wolf-watchers wake up before the sun like monks to point pricey scopes at an active hillside den. Later, tourists rise and submit themselves to the rhythm of Old Faithful's eruptions or hike a mile to give thanks for the kaleidoscopic colors of Grand Prismatic Spring.

These various rituals are infused with shared meaning by their context. They take place in a massive natural cathedral, larger than Rhode Island and Delaware combined, set aside and protected by people who are dedicated to preserving nature in its most pristine form; nature's clergy, so to speak. They are devoted to the earliest scientific mission of Yellowstone, the world's first official national park: "Preserving natural conditions."[7]

The appeal of pure natural conditions—genuine wilderness—has its roots in being the antithesis of the artificial environments where we spend virtually every minute of our existence. "It's something real in this contrived and digital age," argues Doug Smith, a longtime wildlife biologist and the leader of Yellowstone's Wolf Restoration Project. "Life and death. Real nature with no bars in between. Most of us don't get this in our daily lives, so it can be a thirst slaked only by the real thing. There are not many places other than Yellowstone to go for this."[8] Zoos, even the best ones, are simulations. For people like Smith, our "contrived and digital age" is also a simulation, a zoo into which we have placed ourselves. (And who among us hasn't felt this way, at least a little?) But not Yellowstone. Yellowstone is nature, shaped by forces from beyond and before humanity. It is wild, and therefore it is real.

To fully experience the wildness of Yellowstone I decided to camp solo in the backcountry. Vehicles out of earshot; buildings and roads out of sight; animals my only companions. Wasn't that Yellowstone in its purest form? I'd read numerous essays about "the pricelessness of untampered nature," as the ecologist Paul Errington puts it, and I wanted to evaluate it for myself.

Everyone I spoke with approved. Two of my guides, the naturalist Ashea Mills and her husband, ecologist Michael Tercek, told me stories about their own life-changing backcountry experiences. Rangers praised me. "Only one to two percent of visitors even get off the road," one said scornfully, as he helped me find an April hike that would be tolerably free from snow. "You're doing it the right way." When I sat down to talk with Doug Smith, he advised me to take at least a couple breaks from yelling "Hey, Bear!" standard practice for avoiding dangerous encounters when you're hiking alone. What was the point of wildness, he mused, if we are constantly contaminating it with the sound of our own voices?

I knew I wanted to immerse myself in untampered nature, but as I planned the trip I struggled with translating that abstract ideal into practice. It wasn't just my voice that could contaminate nature. There were the propane stove for boiling water, the instant oatmeal and freeze-dried beef stroganoff (slogan: "Savor the adventure!™"), my tent that traded the night sky for nylon, my book about the park and my lantern for reading it before falling asleep. I was an impure pilgrim, clinging to unholiness because I couldn't live without it. Even if I went silent, my very existence would keep shouting "Hey Bear!"

Part of me dismissed these feelings as puritanical nonsense. You can't spoil the natural world with a lantern and a book. Not even the most dedicated nature lovers would think to suggest a Yellowstone backcountry trip in April without a can of bear spray and a subzero sleeping bag. There will always be distance between humans and the rest of nature, and that's just fine. Striving to fully close the gap can lead to a situation like that of Timothy Treadwell, the bear lover and activist immortalized in the Werner Herzog documentary *Grizzly Man*. After Treadwell was devoured by grizzlies in Alaska's Katmai National Park, his partially eaten head, spine, and right arm—watch still ticking on his wrist—were recovered by rangers, a warning to anyone who might be tempted to leave the bear spray at home, as Treadwell did out of respect for his animal companions.

And yet, before his death, Treadwell had managed to cultivate unmediated relationships with multiple bears for over a decade. Was his connection to nature purer than that of the park rangers who insisted on carrying bear spray and installing electric fences around campsites? At a certain point, the security we've come to demand must depend on artificiality and alienation, aerosols and electricity, a distance from natural systems that is at once blessed and tragic. We "unplug" and "disconnect," but not completely. We can't; it isn't in our nature. Even Treadwell slept in a tent.

H. G. Wells captured the paradox beautifully: "Man is the unnatural animal, the rebel child of nature, and more and more does he turn himself against the harsh and fitful hand that reared him."[9] *Unnatural animals.* How then do we act naturally? What are the right rituals?

Martyrdom is the ultimate sacrifice, the fullest cleansing. "I will die for these animals, I will die for these animals, I will die for these animals," said Treadwell, and he did. In the context of the documentary, though, his actions are ambiguous. There's something artificial about the path he has taken, about the religious vision of nature he appears to endorse, the prayers he repeats into his video camera. By the end, I'm with Herzog, whose voice-over forces the viewer to confront the strangeness of Treadwell's religion. "What haunts me," intones Herzog, having reviewed endless hours of footage, "is that in all the faces of all the bears that Treadwell ever filmed, I discover no kinship, no understanding, no mercy. I see only the overwhelming indifference of nature. To me, there is no such thing as a secret world of the bears. And this blank stare speaks only of a half-bored interest in food. But for Timothy Treadwell, this bear was a friend, a savior."

Questions about the right way to experience the park's nature date back to Yellowstone's earliest days. For me, a key part of my experience came before I ever set foot in the park, as I rehearsed a backcountry bear encounter with my wife. She played the mama grizzly, furious that I'd cut her off from her cubs. In this situation, said the internet, you should wave your arms and speak in a soft monotone. "Please do not hurt me, bear," I said to my wife. "I am just a nice human." The bear kept advancing. Time for the bear spray. I popped off the imaginary safety, tested the wind (don't get bear spray in your eyes!) and released a blast. Like approximately 8 percent of actual bears in the wild, this one continued undeterred. I dropped to the floor and lay flat on my stomach as my wife clawed gently at my face. By the end we were laughing, but this preparatory ritual was also a serious reminder about authentic nature. Wild animals are not Disney cartoons. They are not our friends, and some of them are extremely dangerous.

In addition to bear spray, backcountry camping today requires a dedicated bag or bear container to store your food, along with a rope to string it up high in the trees while you sleep. Food waste disposal anywhere other than a designated bear-proof garbage container is prohibited in all areas of the park. Meticulously securing my own food was another ritual performed in honor of pure nature. In pure nature, bears feed on everything from fat-rich army cutworm moths to clover, but never instant oatmeal or freeze-dried beef stroganoff, not even a drop. This isn't just about protecting myself from an unwelcome nighttime visitor; it's also about making sure I am a welcome visitor, a polite guest rather than a disruption. Leave everything cleaner than you found it.

One hundred years ago, preparatory rituals for visiting Yellowstone would have looked much different, because the bears that people encountered were not fully wild. You didn't bring pepper spray to keep bears away; you brought food to attract them. In a 1923 photo, Yellowstone superintendent Horace Albright is seen offering a snack, bare-handed, to one of the park's many black bears that were accustomed to taking treats from humans.[10] In another photo, taken in 1924, he shares a table with three bears who happily dine off plates. Understandably, tourists delighted in doing the same.[11] "The bears of Yellowstone eventually became immortalized in television cartoons as Yogi and Boo-Boo, forever devising new methods of stealing picnic baskets," explains the historian James Pritchard. "In a sense it was classic Pavlovian response: The bears had been trained over many years by the tourists to associate snappy Fords and shiny Hudsons with a food supply, so the result was begging bears by the roadside."[12] In a report that catalogued problems with animal management in national parks, the Wildlife Division of the National Park Service bemoaned how visitors looked forward to seeing a "galaxy of bears at the garbage platform."[13] But it wasn't the tourists' fault. One platform by the Old Faithful geyser had been officially installed with benches meant for tourists, allowing them to sit comfortably during lectures from Yellowstone naturalists while watching the "bear show." After only a few decades, human food waste had become an essential part of the Yellowstone bears' feeding patterns, and watching them eat it a central Yellowstone ritual.

At this time, the image of bears as friendly companions was already well established in the popular imagination. In 1903, Brooklyn candy shop owners Morris and Rose Michtom invented the iconic toy beloved by Christopher Robin and children worldwide: the teddy bear. That winter, the governor of Mississippi had invited President Roosevelt, an avid hunter, to visit his state for a bear hunt. Roosevelt went three days without seeing a bear—unlike their Yellowstone counterparts, Mississippi bears remained wild—so hunt guides took matters into their own hands. They tracked an old black bear wounded by hounds, tethered it to a tree, and invited the president to shoot. Roosevelt refused on the grounds that it would be unsportsmanlike, and newspapers quickly picked up the story. The *Washington Post* ran a cartoon about Roosevelt's refusal, complete with a sympathetic, cuddly-looking bear. Inspired by the story, Rose Michtom made two stuffed bears, and Morris contacted the president for permission to call them "Teddy's Bears," which Roosevelt granted.

Then, in 1926, A. A. Milne ensured the enduring popularity of teddy bears with his invention of Winnie-the-Pooh, paw stuck in a jar of honey like a black bear's paw in an unsecured garbage can. Milne's inspiration came from his son Christopher's stuffed bear, which was named after a real-life tame female black bear, Winnie. Winnie lived at the London Zoo, where Christopher befriended her and visited often, playing with Winnie in her concrete enclosure as Milne looked on approvingly. For children and adults alike, feeding bears in a national park would have represented the natural culmination of a common fantasy.

That the teddy bear, an icon of inaccurate anthropomorphization, should bear Roosevelt's name is highly ironic, given the president's eventual involvement in a campaign against unrealistic representations of nature. Only a few months after that Mississippi hunting trip, the venerable naturalist John Burroughs published "Real and Sham Natural History," a scathing magazine article that blasted popular nature writers for favoring romance over truth in their depictions of wild animals. In the tradition of Aesop's fables, best-selling authors such as Ernest Thompson Seton and the Reverend William J. Long portrayed creatures as if they were human. But Aesop never claimed to be a naturalist. These authors did, and Burroughs was incensed. "No pleasure to the reader, no moral inculcated, can justify the dissemination of false notions of nature, or of anything else," he wrote.[14] Yes, Seton and Long entertained young readers with their stories of animals whose intelligence and cultural complexity were no less than human: a fox that lures hounds onto train tracks, perfectly timed to dispatch them; crows that attend school and are instructed like pupils in a human classroom. But to claim, as these authors did, that their stories reflected genuine natural history was simply unacceptable. "There is nothing in the dealings of animals with their young that in the remotest way suggests human instruction and discipline," insisted Burroughs. Authentic wonders of nature were more than adequate, and did not need the embellishment of fabulists who had the temerity to pass their anthropomorphic fictions off as fact.

Burroughs's essay prompted a fierce back-and-forth about the proper depiction of nature. It raged on for five years, leading to an essay by Roosevelt himself, entitled "'Nature Fakers,'" in which he joined Burroughs to condemn books that deceived readers about the true nature of nature. "The fables they contain," pronounced Roosevelt, "bear the same relation to real natural history that Barnum's famous artificial mermaid bore to real fish and real mammals."[15]

This was no mere literary squabble. The stakes of the "nature fakers" debate were theological. In the words of a *New York Times* article published at the time, it was a referendum on who should "officiate in the temple of Nature and . . . bring to her altars the tributes of truthful and accurate praise."[16] The physical temples of nature were national parks. Just a few months after Burroughs wrote his controversial essay, Roosevelt chose him as a traveling companion for two weeks in Yellowstone, a trip that culminated in the president placing the cornerstone of Yellowstone's famous Roosevelt Arch, which was inscribed "For the Benefit and Enjoyment of the People." At the heart of Roosevelt and Burroughs's displeasure with the nature fakers was the meaning of this inscription. What kind of experience would benefit pilgrims who visit national parks? "Amid the decay of creeds, love of nature has high religious value," wrote Burroughs. "Every walk to the woods is a religious rite, every bath in the stream is a saving ordinance." In nature there is "an inexhaustible field for inquiry, for enjoyment."[17] For Burroughs, enjoyment was linked directly to inquiry, to truth and authenticity. Mythologize nature and the walk is no longer an authentic religious rite; the benefit and enjoyment will be lost.

Just as Burroughs and Roosevelt objected to fake visions of nature in literature, the Wildlife Division of the National Park Service eventually objected to fake scenes of bears gathered at the garbage dump. Nature fakers compromised real knowledge and appreciation; the "galaxy of bears" that tourists had come to expect was no different. And so, in 1932, the Wildlife Division issued a report calling for the "realization that the unique charm of the animals in a national park lies in their wildness, not their tameness . . . an appreciation of the characteristics of a real wild animal, notably, that each

wild animal is the embodied story of natural forces which have been opera-
tive for millions of years and is therefore a priceless creation, a living em-
bodiment of the past, a presentiment of the future."[18] By changing people's
understanding of nature they hoped to change the rituals that were practiced
in nature's temple.

This realization about the unique charm of wildness is now set in stone,
built into the meaning of "benefit and enjoyment." It is taught to children
from an early age and emphasized in park rules and regulations. Feeding
bears in Yellowstone now seems as sacrilegious as shouting obscenities in
church. Our understanding of national parks, and, consequently, the form of
the rituals we practice in them, cannot be separated from the intrinsic value
of wild nature. For this we have visionaries like Roosevelt to thank.

Yet the figure of Roosevelt also serves as a stark reminder that members
of the same faith can practice very different rituals, and that holiness in one
generation may come to be seen as sacrilege in the next. "It is always lawful to kill dangerous or noxious animals, like the bear, cougar and wolf," he
once declared, a sentiment that aligned him with nearly every naturalist and
scientist of the day, all of whom distinguished between "good" and "bad"
animals. Back then, a cleansing ritual could have meant purifying nature of
its less desirable inhabitants. The 1894 Act to Protect the Birds and Animals
in Yellowstone National Park specifically exempted "dangerous" animals,
which included predators such as coyotes and wolves. Even progressive scientists who advocated for the preservation of natural conditions still approved of eliminating rattlesnakes, as well as culling hawks near roads so that
songbirds would flourish.[19]

Ecologists came to reject such value judgments by the mid-twentieth
century, but the general public proved more intransigent, with disastrous
consequences for "noxious" animals. Pure nature still meant purified nature,
not pristine nature. By the 1970s grizzlies were endangered, and wolves—
"beasts of waste and desolation" according to Roosevelt—had long been ex-
terminated from Yellowstone. Their fate was due in large part to the activity
of zealous hunters who, like the president, felt that one of the best rituals
for connecting with nature is killing the creatures that inhabit it, especially
when the world is better off without them. And in a strange twist, the slow
substitution of reverent wolf-watching for righteous wolf-hunting happened
because of anthropomorphic stories like those that Roosevelt and Burroughs
found so problematic.

———————

After decades of working to reintroduce Yellowstone's wolves, the biologist
Doug Smith knows firsthand the visceral hatred and fear they can provoke.
When I told him about my background in religious studies, he smiled grimly
and nodded. "You know," he said, "that makes a lot of sense. There was a
woman once at a local meeting I went to, and she stood up and pointed her
finger at me and started yelling. 'Wolves are not a part of God's world!'
That's what she said. 'All other animals are godly, wolves are ungodly.'"

It may sound extreme, but the woman's statement accurately reflects
the place of wolves in the Western historical imagination for millennia: ungodly, evil, dangerous, deceptive. "Beware of false prophets," admonishes
the Gospel of Matthew, "who come to you in sheep's clothing, but inwardly
are ravening wolves." Shakespeare used wolves repeatedly to represent our
worst attributes: "Thy desires / are wolvish, bloody, starved, and ravenous."
In his classic study *Of Wolves and Men*, Barry Lopez catalogues the myriad
ways that wolves have represented departures from God's natural benevolent order. Precancerous lumps in the breast were known as "wolves" to
seventeenth-century Europeans, as were open sores and growths on their
legs. Famine was "the wolf."[20] Magical werewolves blurred the line between
animals and humans, unholy violations of divinely ordained categories. If
Christ is a shepherd, argued one nineteenth-century Bible commentary,
then Satan is a wolf: "1. His attacks are deadly. 2. His surprises are crafty.
3. His hatred of Christ is implacable. 4. His hunger to devour is insatiable.
5. He attacks under darkness. 6. He scatters the flock by tempting them to
luxury, avarice, and sensuality. Filling their minds with pride, envy, anger,
deceit."[21] The Big Bad Wolf, hungry for the flesh of innocents.

The sordid history of wolf-killing in North America is that of a crusade
against evil. At the turn of the century, some hunters would take live captured wolves and cut off their lower jaws or wire their mouths shut, then release them to die a slow death by starvation. When you are saving Little Red
Riding Hood and her grandmother, it's not enough to kill the wolf—you
must dump rocks into his stomach and sew it back up, ensuring he awak-
ens to a gruesome punishment. Exterminating wolves was understood as
a salvific mission, literally and figuratively. Midwestern soldiers back from
World War II referred to wolf-hunting as "killing Nazis."[22] Jay Hammond,
the future governor of Alaska, wrote a 1955 article for *Field & Stream* in

which he boasted that his elimination of 259 wolves by aerial strafing—a technique that is now illegal—had saved the lives of Eskimos, who would have otherwise struggled to compete for game. (The Inupiat Eskimos themselves eventually complained about the scarcity of wolves and noted that caribou were no more abundant than before the massacre.)[23]

"Part of this is reasonable, and part of it isn't," Smith said, when I asked him about fear of wolves. He readily acknowledged that wild wolves kill livestock. But he also expressed exasperation at some ranchers who inflate the numbers, and complained about their apparent unwillingness to look objectively at livestock depredation figures. Negative attitudes toward wolves, in his experience, go far deeper than the harms they inflict. The novelist and nature writer Brenda Peterson remarks that wolf preservation is often referred to as "the abortion issue of wildlife," a metaphor that gestures to its religious undertones, and the extent to which people's identity is profoundly bound up with their position: pro-wolf or anti-wolf.[24]

"Wolves are the opposite of American cowboys," Smith explained to me. "When we were moving west, wolves killed cows, so we shot as many as we could. Wolves are a symbol of wild nature, which is exactly what we were trying to conquer. Now people still hate them for what they represent."

That the pro-wolf camp eventually succeeded in reintroducing wolves to Yellowstone speaks to a massive shift in the public's understanding of wolves, and nature more generally. A significant part of that shift was scientific. Overwhelming evidence showed that the population level of a specific animal in an ecosystem could not be radically altered without far-reaching consequences. It was no longer scientifically coherent to conceive of some creatures as "noxious" or "ungodly." To "purify" nature of wolves was in fact to pollute the whole. If ecosystems are God's creation, then every creature is of God's world, not just the ones that please us.

Along with the scientific change in perspective came powerful new stories about wolves. Perhaps the most important of these was "Lobo, the King of Currumpaw," published in 1898 by none other than one of the vilified "nature fakers," Ernest Thompson Seton. Before his career as a writer Seton had been a wolf-hunter, and in "Lobo" he recounts his experience hunting an old gray wolf in New Mexico. Lobo epitomized everything people hate in wolves: "All the shepherds and ranchmen knew him well," recalls Seton, "and, wherever he appeared with his trusty band, terror reigned supreme among the cattle, and wrath and despair among their owners."[25]

Seton set out to kill the monster, but time and again Lobo eluded his poison and traps with "diabolic cunning." Finally he managed to kill Blanca, Lobo's mate. But what should have been a triumph ended up a tragedy. The King of Currumpaw howled plaintively, grieving the loss of his queen, and Seton was overcome with pity and regret:

> "Blanca! Blanca!" he seemed to call. And as night came down, I noticed that he was not far from the place where we had overtaken her. At length he seemed to find the trail, and when he came to the spot where we had killed her, his heart-broken wailing was piteous to hear. It was sadder than I could possibly have believed. Even the stolid cowboys noticed it, and said they had "never heard a wolf carry on like that before." He seemed to know exactly what had taken place, for her blood had stained the place of her death.[26]

It's easy to imagine Burroughs arguing that wolf hearts do not break; that wolves do not have kingdoms or subjects or queens; that Seton (deliberately) mistakes instinct for cunning and habituation for love. Yet it is Seton's admiring, anthropomorphic language that now dominates popular accounts of wolves, even among hard-nosed scientists. Their paradoxical attitude manifests itself in the jarring contrast between wolf names and descriptions of their character. In Doug Smith and Gary Ferguson's account of reintroducing wolves to Yellowstone, *Decade of the Wolf*, we read about Wolf 106, "highly capable of leadership," a "tireless and . . . benevolent leader of the Geode Creek Pack."[27] There's Wolf 302, who becomes the most "reproductively successful wolf in all of Yellowstone" by "abandoning the usual monogamous relationship approach, typical among wolves, for the life of a rolling stone."[28] The jacket copy on Nate Blakeslee's *American Wolf* offers a deeply humanizing portrait of Wolf 06, one of the book's two central protagonists. She's "a kind and merciful leader, a fiercely intelligent fighter, and a doting mother."[29] Wolves are said to have their own culture, their own civilization. "Everything I see in a wolf's behavior is just like how you'd expect a person to behave," observes wolf advocate Oliver Starr.[30] (Similar tendencies are seen in the discussion of most animals: at the Cincinnati Zoo, for instance, the placards explaining species behavior are frequently titled *"Just like us!"*)

When I watched wolves in Yellowstone I saw them through the lens of these humanizing narratives. My guide Ilona admitted that even though people try to remain aloof, everyone has their preferences. It took little time for me to develop my own preferences as I witnessed an extraordinary social interaction. Near the den we located the breeding, or "alpha," female, as well as a few other wolves. Something was wrong, though—her packmates seemed unwelcoming. Or were they playing with her? As they disappeared behind a hill it was hard to tell; then she reappeared alone, limping badly. Experienced wolf-watchers around me discussed potential explanations. Ilona told me this particular wolf had been excessively dominant over the rest of the pack. Maybe they'd simply had enough? I found myself pitying her regardless, angry at her abusive companions. I wanted a better reason for their behavior.

"Let me ask Rick," said Ilona. "If anyone knows, he will." Together we approached an older man who was watching the pack from his truck. It was Rick McIntyre, one of Yellowstone's most dedicated wolf-watchers, and, alongside Wolf 06, the other protagonist of Blakeslee's *American Wolf*. McIntyre had no explanation for what we were seeing right then, but it didn't matter. To speak with him about wolves in general was to be drawn into epic narratives, ancient rivalries between warring tribes, courageous heroes and vanquished foes. A local newspaper described his accounts as Shakespearean; the author and wildlife activist Rick Lamplugh calls them sermons. This dramatic approach, with anthropomorphization at its center and salvation as its goal, is directly descended from Seton, whom McIntyre regularly praises. Like Seton, he, too, has been criticized for preaching sermons that liberally attribute human traits to wolves, but it hasn't stopped him. "Rick's dream," writes Blakeslee, "though he seldom described it as such, was to someday tell a story so good that the people who heard it simply wouldn't want to kill wolves anymore."[31] With new stories come new rituals. This is the transformative hope: if you see wolves the way McIntyre sees them, the way Seton saw them, then you might trade your rifle for a spotting scope.

Like wolf-hunters, wolf-watchers conceive of their activity in quasi-religious terms. McIntyre refers to wolf-watching as a "peak life experience"—he once went 891 straight days without missing a sighting. Watchers will gush over something as trivial as spotting fresh wolf urine on a snow pole. "Everything is sacred for a true wolf watcher," writes Rick Lamplugh in his account of Yellowstone's wolves, *In the Temple of the Wolves*.[32] The animals are sacred, not only because they are seen as kindred spirits to humans, but also because they fit into a larger redemptive narrative of nature restored, a so-called keystone species whose presence unlocks the true being of Yellowstone. "The only way we can experience 'an entire heaven and an entire earth' is to bring the wolf back," said McIntyre, quoting Thoreau, in the epilogue to his own book, *A Society of Wolves*.[33] Alongside the wolves' worshipful devotees, I felt certain about the righteousness of restoring what our bloody attempts at purification destroyed. Here, at least, was a sliver of nature redeemed. The ritual had fulfilled its purpose.

Part of my guided Yellowstone tour was a visit to a Native American hunting blind, a subtle pile of rocks that once hid hunters from approaching game. Likely built by the semi-nomadic Sheepeater tribe who called Yellowstone their home, the blind was located more than a mile from any designated hiking trail, unknown to all but a few rangers and naturalists. As I followed Ashea toward it through the unmarked grassland, she turned and saw me walking in her footsteps. "Try not to step where I do," she said sternly. "Find your own path."

Finding my own path meant obeying the second principle of the Center for Outdoor Ethics' Leave No Trace Seven Principles: "In pristine areas: Disperse use to prevent the creation of campsites and trails." The principles were developed to help people "enjoy our natural world in a sustainable way that avoids human-created impacts."[34] In Spanish, Leave No Trace becomes Sin Dejar Huellas, literally "Without Leaving Footprints." In Yellowstone, every step I take while mindful of this ethic is tacit acknowledgment that wilderness, pristine nature, is incompatible with even the most basic of human activities.[35] We make it pure not just by replacing wolves, but also by removing ourselves.

The Leave No Trace understanding of wilderness reflects the famous definition enshrined in the Wilderness Act of 1964, which foregrounds the absence of humans: "A wilderness, in contrast with those areas where man and his own works dominate the landscape, is hereby recognized as an area where the earth and its community of life are untrammeled by man, where man himself is a visitor who does not remain."

Our crimes against nature earned us this excommunication from the earth and its community of life. Couches, handkerchiefs, diapers, and, now, drones have all been retrieved from Yellowstone's geysers and delicate geothermal formations, where visitors and park employees once washed their soiled clothing. We gave bears so much of our garbage that after it was withdrawn they struggled to feed themselves. We killed all the wolves and very nearly all the bison before they received legal protection from our own worst instincts.

When we arrived at the Sheepeaters' hunting blind, however, I was struck by the bitter irony of the Leave No Trace ritual and the definition of wilderness that it implies. Long before settlers "discovered" the pristine lands that are now protected from humans, other humans already lived there, not as visitors, but as active members of the community of life. "We did not think of the great open plains, the beautiful rolling hills, and winding streams with tangled growth as 'wild,'" wrote Chief Luther Standing Bear in 1933. "Only to the white man was nature a 'wilderness' and only to him was the land 'infested' with 'wild' animals and 'savage' people. To us it was tame. Earth was bountiful, and we were surrounded with the blessings of the Great Mystery."[36]

The hunting blind highlights an entirely different set of ritual practices that defined the relationship of indigenous people to the natural world. For Chief Luther Standing Bear it would have been bizarre to see humans as passing "visitors" in the wilderness, much less to elevate that role as the best way to experience nature. In his view, human animals make their homes from natural materials. They obtain their food from nature. They hunt just like other predators, borrowing techniques from fellow skilled hunters: bears, lions, wolves, cats, and eagles. But humans are no "better" or "worse" than any other animal. Predator and prey are members of an interdependent community. Animals (and plants) take their place in the world alongside humans as family members with spiritual status, to whom we owe reverence and gratitude. (The word "ecosystem" may have been coined in 1935, but the concept obviously existed long before that.)

As a ritual, Leave No Trace makes no sense for members of such a community. What became the greatest sin in modern national parks—killing wildlife—is for many Native Americans a crucial way of reaffirming their place in the universe, laden with significance. Yes, nature should be respected, but that does not necessarily require an artificial separation of humans from the natural cycles they seek to respect. The hunting blind left a trace that does not make the landscape any less pure. "There are numerous sites in the Yellowstone area that remind us that our ancestors lived and interacted with every inch of this sacred landscape for thousands of years," says Francine

Spang-Willis, an anthropologist and Pawnee woman of the Northern Cheyenne Nation. "Artificial boundaries may exist on the land, but we still have a deep connection to all of it. We still live and interact with it."[37]

Nearing the hunting blind, Ashea and I had to pause and wait for some bison to move away. They stood near the rock pile, living testimony to the need for artificial boundaries, which apply simultaneously to geography and behavior. In the park, hunting is strictly prohibited because certain humans proved manifestly incapable of harmonizing with nature. These humans completely eradicated Native Americans from Yellowstone by 1879, and slaughtered hundreds of thousands of bison in the service of that goal. Professional hunters killed bison with bullets supplied by the US Army's anti-Indian campaign. Throughout most of the nineteenth century, the bison and the "savages" who depended on them were antithetical to the dominant American view of wilderness as something to be conquered by civilization and Manifest Destiny.

The results were devastating. John Fire Lame Deer, a Lakota medicine man, wrote gut-wrenchingly about what the bison meant for his people, and what happened when they were destroyed:

> The buffalo gave us everything we needed. Without it we were nothing. Our tipis were made of his skin. His hide was our bed, our blanket, our winter coat. It was our drum, throbbing through the night, alive, holy. Out of his skin we made our water bags. His flesh strengthened us, became flesh of our flesh. Not the smallest part of it was wasted. His stomach, a red-hot stone dropped in to it, became our soup kettle. His horns were our spoons, the bones our knives, our women's awls and needles. Out of his sinews we made our bowstrings and thread. His ribs were fashioned into sleds for our children, his hoofs became rattles. His mighty skull, with the pipe leaning against it, was our sacred altar. The name of the greatest of all Sioux was Tatanka Iyotake—Sitting Bull. When you killed off the buffalo you also killed the Indian—the real, natural, "wild" Indian.[38]

The tragedy of the dwindling herds started appearing in magazine articles and political cartoons, exciting enough popular concern that Congress introduced several resolutions and bills to protect them from "indiscriminate slaughter and extermination." One bill, progressive for its time, would have made it illegal for non-Native Americans to wound or kill female bison of any age, and illegal to kill male bison for any purpose other than food or hide collection—that is, no wanton slaughter.

Resistance to the bill was strong. Even if people sympathized with the bison, fewer sympathized with Native Americans (the bill "gave preference to the Indians," complained one representative), and the eradication of bison was seen as a necessary step in the elimination of Native Americans. "There is no law which a Congress of men can enact, that will stay the disappearance of these wild animals before civilization," declared US representative Omar Conger in 1874. "They trample upon the plains upon which our settlers desire to herd their cattle and their sheep. There is no mistake about that. They range over the very pastures where the settlers keep their herds of cattle and their sheep to-day. They destroy that pasture. They are as uncivilized as the Indian."[39] When Congress did manage to pass protective legislation in that year, Ulysses S. Grant vetoed it—only two years after signing Yellowstone into law.

When bison finally received legal protection their numbers in Yellowstone stood at a scant two dozen. To remedy the dire situation, a dedicated breeding program was instituted in 1902 at the park's Lamar Buffalo Ranch with the goal of bringing the population back to a sustainable level. By 1945 there were 1,300 bison roaming free, and now the figure stands somewhere between 2,300 and 5,500 depending on the year and the season.[40] Though bison are common as livestock, very few roam wild on public lands, and most of these have cattle genes from interbreeding. The bison I saw in the park are considered exceptional emblems of true nature, barely saved from the humans who would have destroyed them. "Yellowstone is the only place in the United States where bison have lived continuously since prehistoric times," states the National Park Service Yellowstone website's FAQ page with unconcealed pride. "A number of Native American tribes especially revere Yellowstone's bison as pure descendants of the vast herds that once roamed the grasslands of the United States. The largest bison population in the country on public land resides in Yellowstone. It is one of the few herds free of cattle genes."

As the bison finally ambled away from the hunting blind, it struck me that one day some of them would cross the park's boundary and, like their ancestors, die at the hands of eager human hunters. Yellowstone's ecosystem may be close to natural, but the geographical and behavioral boundaries that define it are artificial, including the prohibition on hunting. And so, every year, hunters enter a lottery overseen by the Montana Department of Fish, Wildlife and Parks, in hopes of winning a chance to kill Yellowstone bison. Demand is extremely high—according to statistics from 2015, over ten thousand hunters vied for only seventy-two licenses, which is fairly typical.[41] The lucky state hunters are joined by tribal hunters, Native Americans who receive permits through their respective tribal governments, the fulfillment of a long-neglected 1855 treaty with the US government. These permits are also in high demand: when the Blackfoot Nation joined the hunt in 2018, eighty permits sold out almost immediately.[42]

Hunting season is in winter, when bison migrate out of Yellowstone proper. Their behavior is entirely natural, which explains why the back wall of Doug Smith's office does not feature a map of Yellowstone, but rather a map of the Greater Yellowstone Ecosystem, an area nearly ten times the size of the park. Scientists like Smith understand that Yellowstone's borders were originally established with little regard for, or knowledge of, the natural systems that run through them. The exigencies of winter dictate that groups of these shaggy giants must leave Yellowstone in search of food. Most of them cross the border in Beattie Gulch, a tight migration corridor just north of the park. That's where state hunters and tribal hunters wait, rifles ready for the moment when, like magic, the unknowing bison's political protection disappears and they become fair game.

For state hunters, killing a bison represents the fulfillment of a rare opportunity immortalized by Roosevelt himself, who wrote ecstatically about shooting a bison in 1893: "He was a splendid old bull, still in his full vigor, with large, sharp horns, and heavy mane and glossy coat; and I felt the most exulting pride as I handled and examined him; for I had procured a trophy such as can fall henceforth to few hunters indeed."[43]

For Native Americans, it means something wholly different. They are participating in an ancient, sacred ritual, once central to the subsistence of indigenous peoples. "I go there fulfilling the legacy of my ancestors,"[44] said Nez Perce tribal member James Holt, who was instrumental in securing the rights of Native Americans to hunt bison. Jeffrey Sampson, a member of the Yakama tribe, echoed Holt when explaining why he joined the hunt in 2018: "We're a dying generation and we try to preserve what things we can that's valuable amongst our culture."[45]

Yearly protests demonstrate that not everyone agrees with the characterization of the bison as "fair" game, or even with dignifying the event as a hunt. "Those people . . . some in camo, some in bright orange vests—they're not hunters," writes Lamplugh, reflecting on his experience watching the killing of twenty-one bison. "They are shooters—a firing squad, really."[46] My guide Ashea recalled with horror watching multiple bison shot at point-blank range. "It was absolutely horrific," she told me. "A bloodbath."

The animals live a cloistered existence in their sanctuary where humans gather only to admire them, so they have no instinctual sense of the danger they face upon leaving. Like the tourists who mistook tame bears for the real thing, the bison have become accustomed to tame humans, artificially pacified by Yellowstone's regulations. They are unfamiliar with humans as the most effective predators on earth.

Debates over the ethics of hunting bison are particularly controversial when they involve condemnation of Native Americans. In 2017, the Buffalo Field Campaign, a conservation group, released photographs of what appear to be teenage Nez Perce hunters stabbing a bison that had been shot but not killed. That same year, the Montana Department of Fish, Wildlife and Parks cited multiple instances in which officials were forced to put down bison that had been wounded by inexperienced hunters and left to die. Tribal representatives issued a swift response. The hunt, they argued, was being conducted lawfully, and tribe members who violated laws had had their licenses revoked. "To me, it's like they're trying to portray us like we're savages or something," said McCoy Oatman, the vice chairman of the Nez Perce Tribal Executive Committee, in an interview with the *Bozeman Daily Chronicle*.[47]

Nevertheless, some activists insist that all arguments in favor of hunting, indigenous or otherwise, are irredeemably flawed. In her scathing essay "The Killing Game," Joy Williams dissects the horror of hunting whales whose existence is as cloistered as that of Yellowstone's bison:

> The subsistence/cultural/traditional argument is as egregious as it is canny. The Makah Indian tribe of Washington won the right to kill the gray whale in 1999 despite the U.S. ban on whaling, and since the whales in Neah Bay on the Olympic Peninsula were used to being approached by tourist boats, a successful hunt was pretty much assured. A young whale was easily killed with high-powered rifles and hauled ashore with the assistance of some commercial fishing boats, but the community hadn't a clue as to what to do with the two tons of meat that resulted, blubber not having been a part

of the tribe's diet for over seventy years. After dancing on the dead whale's back for a while, everybody went home, leaving the meat to rot. Granted that there might be a few "good" hunters out there who conduct the kill as spiritual exercise and a few others who are atavistic enough to want to supplement their Chicken McNuggets with venison, most hunters hunt for the hell of it.[48]

Though her writing has an air of finality, Williams's passing reference to "good" hunters allows that hunting might not be intrinsically bankrupt. If you conduct the hunt ritually, as a spiritual exercise, if you eat what you kill, if there's some chance of your prey escaping—in those cases the act of pulling the trigger could be redeemed.

I've never hunted, and before visiting Yellowstone I shared Williams's skepticism about the existence of such hunters. Surely the desire to personally kill an animal is more likely to be rooted in a love of power than a love of nature. Perhaps Native Americans were once part of a natural system, hunting on foot with bows and arrows out of genuine necessity, but now they pull up in the same pickup trucks as the state hunters, wear the same outfits, fire the same rifles, participate in the same ritual that was responsible for the death of the "real, natural, wild Indian" that John Fire Lame Deer regrets.

That's why I was surprised when Doug Smith told me he hunts elk—and, unlike the wolves, who hunt to survive, for him it's a choice. ("Once the freezer's full, we stop," he told me.) More surprising still, I found out that the Buffalo Field Campaign, the most vocal bison rights organization in the country, occasionally feeds its volunteers with elk meat that has been hunted by other members.

Spend just a few hours talking with residents of Bozeman, and it is impossible to maintain the position that hunting is necessarily opposed to a love of nature. "My neighbor hunts grouse, and elk, and wolves," Ashea told me. "You know what I think he loves? Tromping through the woods. He's a really good land steward. I know he'd love to bag a wolf, but he also came to me distraught because he found pups in an abandoned wolf den and wanted to save them." These hunters are the complicated descendants of Roosevelt, and of John Fire Lame Deer, working tirelessly to preserve the natural world while seeking to connect with it through the act of taking life.

Among the "good" hunters there are still bitter disagreements. In *To Save the Wild Bison*, Mary Ann Franke notes that even the concept of "fair chase" remains controversial. Giving animals "a sporting chance" is central to most nonindigenous hunting ethics, including the mission of America's oldest hunting club, Boone and Crockett, founded by Roosevelt in 1887. Yet the InterTribal Bison Cooperative rejects as Eurocentric any view that "associates the difficulty of killing an animal with fairness and a belief that merely shooting the animal is unacceptable." According to a statement from the cooperative, "the important distinction to Indian people is not between hunt and shoot, but between taking an animal's life with or without respect."[49] From this perspective, shooting a bison point-blank looks less horrifying.

Though the conservationists I spoke with differed dramatically on the ethics of hunting bison, there was one notable point of agreement: an ideal of hunting "naturally" is irrelevant. No one suggested that Native Americans who want to hunt bison should be forced to wear moccasins and use bows and arrows. For those who oppose hunting in any form, it does not matter that human beings are "natural" hunters—what matters is whether hunting causes unnecessary pain, or threatens the well-being of a species. The rituals that bring us closer to animals, that honor animals, need not be natural themselves. It's a seeming contradiction that dissolves as soon as naturalness ceases to be synonymous with holiness.

My own ritual visit to Yellowstone removed any lingering suspicions that the idea of nature is a "social construct." Naturalness is a continuum, and it can be very difficult to decide where something exists on that continuum. It may be the case that human influence has spread so far that nothing is purely natural anymore. But there is no denying that certain spaces and systems are less touched and less ordered by humans than others. Walking alone through the grass, immersed in one of those nearly untouched systems, I was fully convinced that nature has intrinsic value and deserves protection, just like a historical monument or a classic piece of art. Resources spent on saving the bison, on bringing back the wolves, on carving out cathedrals for nature—these are wise decisions that make the world a better place.

Yet as I gathered my waste and placed it in the nylon bear bag for the last time, I also felt certain that, paradoxically, my ritual visit was undiminished by its unnaturalness. You don't need to be a "grizzly man" to understand and care for grizzlies. You can take the approach of a scientist and tag them, collar them with radios, study them from airplanes, crunch data about their eating habits on a computer. You can observe them through a scope, or from an automobile. You can connect with elk by photographing them; you can integrate yourself into an ecosystem by shooting them. To honor nature there's no need to remake ourselves in its image, an impossibility for unnatural animals like ourselves. Instead, we can do what wolf-watchers do after a glorious morning; what Native American hunters do after successfully felling a buffalo; what my guides do every day as they look up at ancient mountains and down at new-formed soil. I learned another ritual from these diverse lovers of nature, an alternative to being natural.

It is giving thanks.

CHAPTER 5

Let Food Be
Thy Medicine

*Many people have found that as God's presence and love
become more real and tangible, healing and regeneration
occur naturally.*

—OFFICIAL CHRISTIAN SCIENCE WEBSITE

THE HIPPOCRATES HEALTH INSTITUTE is a natural healing retreat set in
the lush tropical landscape of West Palm Beach, Florida. Palm-lined walking
paths link up a complex of beautiful beige stucco buildings with Spanish-
style roofs. You could easily mistake it for an upscale resort, if not for the
signs that line the roads: *UNLOCK YOUR ENERGY. START LIVING
YOUR BEST LIFE . . . NOW. LIBERATE YOUR VITALITY THE
FUTURE OF HEALTH IS HERE.*

In 2014, the institute and its director, Brian Clement, featured in a tragic
Canadian legal battle. Two unrelated First Nations girls from Canada, ten-
year-old Makayla Sault and eleven-year-old J.J. (a pseudonym), had been
diagnosed with acute lymphoblastic leukemia. Their respective parents
eventually chose to withdraw them from grueling chemotherapy regimens,
opting instead to send the girls to HHI for its unique natural therapies, introduced to them by Clement while he was touring Canada.

The Canadian government has a terrible history of trampling on First
Nations culture and rights, mistreatment that includes wresting children away from parents and handing them to the state. With that history
looming in the background, child protective services elected not to inter-
vene in Makayla's case. Her family raised and spent $18,000 on Clement's

107

"comprehensive cancer wellness program," a combination of vitamin C injections, vegan raw food diet, wheatgrass shots, and "cold laser therapy." Once given a 75 percent chance of survival by oncologists, Makayla died less than a year after she first made headlines.

J.J.'s situation was similar, but in her case the hospital went to court in an attempt to force her parents into continuing chemotherapy. The judge ruled in favor of the parents over the hospital, citing their protected constitutional right to pursue traditional medicine. Like Makayla, J.J. relapsed after discontinuing chemo and attending Hippocrates. However, the relapse led her parents to reconsider their decision, and eventually they concluded that "chemotherapy, along with traditional Haudenosaunee (Iroquois) medicine, which J.J. had already been receiving, would be the next best step." Unlike Makayla, J.J. survived.[1]

Comparing HHI's approach with traditional Haudenosaunee medicine, it's hard to understand why it would appeal to the parents. No First Nations healing tradition has ever emphasized a raw diet or veganism, much less intravenous vitamin C and cold lasers. Many of the institute's therapies have names that invoke the authority of cutting-edge Western science, not ancient indigenous wisdom: the OndaMed Pulsed Electromagnetic Field Therapeutic System, the NuCalm System of Clinically Proven Neuroscience Technology, and Deepak Chopra's "Dream Weaver" Neuroscience Technology. Wheatgrass juice, which is part of every Hippocrates treatment plan, was first researched in the 1930s by Charles Schnabel, an American agricultural chemist, and popularized for healing and detoxification a decade later by Ann Wigmore, the Lithuanian-American founder of the Hippocrates Health Institute. The bright green juice is made from common wheat, a plant introduced to the Americas by the same colonizers who wiped out indigenous populations. Those therapies on offer that do appeal to the ancient wisdom of non-Western traditions—yoga, ayurveda, acupuncture, mandala drawing—come largely from India and East Asia.

Clement himself only complicates the mystery. When I met with him he looked exactly as he does in his online videos: dress shirt and tie, slicked-back dyed hair, trim goatee, gleaming white teeth, eerily tan skin—the living caricature, really, of a snake-oil salesman. On his desk were three bottles: cayenne pepper extract, a disinfectant, and something he identified as "oxygen." Throughout our conversation Clement never brought up indigenous healing practices or traditions, choosing instead to emphasize the financial corruption of Western medical culture, the arrogance of academic science, and the remarkable successes of the Hippocrates program in reversing disease. The only evidence of Clement's connection to First Nations peoples was an exquisitely carved stone bust of an elderly woman, head resting on an ear of corn, prominently displayed in the center of his office: a gift from "one of the girls' families," who felt badly that he had been blamed for Makayla Sault's death. The statue, he explained, meant they did not hold him responsible, unlike biased members of the media. It meant the girls' families stood behind what he knew to be the real story—the hospital, not Hippocrates, had killed her.

Because of liability issues, in public interviews Clement repeatedly insists that his approach neither "cures" nor "heals." But after we finished talking, he encouraged me to buy a documentary called *Healing Cancer*, prominently displayed at the institute's store, with an unambiguous tagline: "If you think conventional treatments cure cancer . . . If you think diet cannot cure cancer . . . Think again."

How did it happen? How did these First Nations families come to trust Clement, despite his distance from their culture, his ignorance of their medical traditions, and what looked to me like the most obvious signs of charlatanry? Without visiting the institute it's almost incomprehensible. Yet if you meet the diverse clientele in person, if you talk with them about "living foods" and watch as they juice their own wheatgrass, if you walk with them along a palm-lined path to the organic greenhouse where the wheatgrass is grown on-site, the beginnings of an answer emerge. Hippocrates provides a fundamentally different approach to healing than what's on offer in hospitals, an alternative theory of disease, health, and the body, and a corresponding set of rituals that stand in stark contrast to the scientific treatments of modern medicine.

These rituals are as diverse as the clientele that practices them, but they are united by their purported connection to the power of natural healing— the same power that somehow makes cold laser therapy a kindred spirit of Haudenosaunee medicine, the same power that Ann Wigmore distilled into her books (such as *Overcoming AIDS and Other "Incurable Diseases" the Attunitive Way Through Nature*) and her bright green juice, the nourishing blood of Mother Nature.

There are countless reasons for skeptics to rebel at the very idea of natural healing. It's a sprawling category with no clear criteria that encompasses everything from raw food to infrared saunas to electromagnetic pulse machines. Browsing the supplements at Whole Foods can feel like time-traveling to a medieval apothecary, the word "natural" on the labels conferring its mystic blessing to "immunity boosters" and "brain optimizers." In the same aisle you'll find homeopathic remedies, evidence that belief in the power of nature is associated with active rejection of basic scientific consensus—harmless, maybe, when it comes to taking inert sugar pills, but dangerous when it means rejecting vaccines. Worst of all, there are the tragedies that happen under the auspices of natural healing: the people who suffer and die, the family members who spend ruinous amounts of money on rituals that do nothing more than summon up false hope.

However, reducing natural healing to these damning examples is a mistake, not unlike the mistake Clement makes when he characterizes "modern medicine" almost entirely in terms of corrupt pharmaceutical companies and arrogant physicians. Natural medicine as we know it today dates back to the nineteenth century, and its emergence is intimately related to millennia of debates about the proper relationship between healing and science, body and mind, and, most importantly, doctors and patients. These debates continue into the present, for despite the great virtues of science, they cannot be settled with laboratory experiments or randomized controlled trials. Health is not fully reducible to physiology. Illness can feel like your body and the world are betraying you, and therefore healing (and healers) must deal with the existential trauma of that betrayal along with its physical causes and symptoms. That trauma is difficult to measure. It does not reveal itself in CT scans and blood tests. But the existential betrayal of illness is no less real or painful than a broken arm, and for thousands of years healing was not complete without ritually addressing it.

Proponents of natural healing regularly integrate ancient medical traditions into their approach, reverently citing the teachings of Hippocrates and the holistic wisdom of classical Chinese medicine. However, these early medical traditions had radically different understandings of what "natural" meant. The familiar distinction between natural healing and other forms of healing—typically pharmaceuticals and surgery—is largely a product of the last two centuries, and would have made little sense to traditional healers and the public to whom they ministered.

Until recently, most cultures recognized three different illness etiologies: supernatural causes (gods, demons, spirits), magical causes (sorcery), and natural causes (arrow wound, rotten food). Any appeal to traditional healing is incomplete without recognizing the causal roles of the supernatural and the magical. According to the Tohono O'odham Native Americans of the Sonoran Desert, for instance, illness can result from breaking a ceremonial taboo, incurring the ill will of an animal, or having a foreign object in one's body that was sent there by a sorcerer.[2] Homer's *Iliad* famously opens with Apollo raining pestilential arrows down on the Achaean forces, spreading plague among the men. And in Shang Dynasty China (ca. 1500–1046 BCE), disorders were frequently ascribed to curses from deceased ancestors. Royal diviners diagnosed the king's maladies by inscribing turtle plastrons and ox scapula with inquiries, as in the following examples:

Tooth illness. Is there a curse? Perhaps from the deceased father Yi?

The king is ill. Was he perhaps cursed by the deceased grandmother Chi? Or the grandmother Keng? Will the condition become serious?

Swelling of the abdomen. Is there a curse? Does the deceased Chin-wu desire something of the king?[3]

Healers employed therapeutic methods that corresponded to these causes: appeals to the gods, magical talismans and formulas, and medications. To placate Apollo and stop the plague, the Achaeans returned an enslaved woman to her father, one of Apollo's priests, hoping the god would have mercy. But to heal an arrow wound, high-ranking Achaean soldiers might have visited the great military surgeon Machaon, who would carefully cut the arrow out and rub the affected area with ointments. Ancient Egyptians had numerous gods, even more spells, and a vast pharmacopeia that included ingredients ranging from honey and pomegranate juice to lead salts and human excrement. (If you were bald, a classic medical papyrus recommended treatment with extracts from "female vulva, male penis, and black lizard.")[4] People commonly wore magical protective amulets to ward off illness. But if the amulet failed, you might visit a priest who could ritually beseech your favored god, or ask a doctor for medicine.

Despite the theoretical utility of categories such as "natural" and "supernatural" for modern scholars, in practice they are almost always blurred. A supernatural force can use a natural vehicle to harm people, as Yahweh did to the residents of Ashdod when he afflicted them with tumors. The tumors were understood to be divine punishment for the Philistines' theft of the ark from the Israelites, and curing them required its return, along with an additional ritual offering of five gold tumors and five gold rats. "The episode," explains historian of religion and medicine Gary Ferngren, "indicates that the Philistines could posit a causal relationship between rats and disease, while at the same time attributing the plague to the anger of Yahweh."[5]

Similarly, what looks like magical causation today—sticking pins into an effigy, for example—makes use of what were once considered objective principles built into the natural world. In the case of harming an effigy, the relevant principle was "like causes like." Supernatural, natural, magical—when it came to identifying the cause of an illness, they all overlapped. The traces of these common origins remain in language. In German, "to heal" is *heilen* and "holy" is *heilig*. In Spanish, "cure" is *cura*, which is also the word for "priest."

Ancient forms of medical intervention also defy easy categorization. Hermes, a god, gave Odysseus a special herb, "moly," that granted him resistance to the "evil drugs" (*kaka pharmaka*) employed by Circe to trap his men. Though a plant, moly is described as "dangerous for a mortal man to pluck from the soil, but not for the deathless gods": a natural protective agent, granted to humans through supernatural intercession. The Greek word for medicine—*pharmakon*—itself blurs the line between magic and nature, since its many meanings include not only "medicine" but also "charm" and "spell." As for the distinction between magical and religious healing? The scholar of myth Joseph Campbell liked to say that myth is just other people's religion, and in that sense, magic can be thought of as other people's religious rituals. For this reason, historians of early medicine employ hybrid terms such as "magico-religious" to more accurately represent the practices they are describing.

Definitional difficulties aside, scholars generally agree that two of the most important early medical texts, the Hippocratic corpus in Greece (ca. 400 BCE) and the *Yellow Emperor's Classic of Medicine* in China (ca. 300 BCE), represented a crucial turn to naturalistic medicine. Since places like the Hippocrates Health Institute enthusiastically endorse Hippocrates and ancient Chinese medicine, it would be easy to confuse "naturalistic" medicine with an early version of today's "natural" healing. In fact, the term "naturalistic" refers specifically to the rejection of supernatural etiologies and therapies. "One who is enthralled with demons and spirits cannot be spoken with," admonishes the

Yellow Emperor's Classic of Medicine, which includes no magico-religious formulas for exorcisms or spirit appeasement. Likewise, the Hippocratic treatise on epilepsy, *On the Sacred Disease*, asserts that epilepsy is not "more divine or more sacred than other illnesses." Against conventional medical wisdom of the day, the author argues that epilepsy "has a natural structure and rational causes," warning the reader about those who say otherwise:

> In truth I maintain that the first to confer a sacred character to this disease were men the likes of which are still around today, magicians and purifiers and charlatans and impostors. . . . They took the divine as shelter and a pretext for their own incapacity—since they did not know with what therapy they could give benefit—and so as not to manifest their total ignorance, they asserted that this disease was sacred.[6]

In the ancient Greek context, "natural" interventions would not have made sense as something in binary opposition to surgery or manufactured drugs. To the contrary: surgery and drugs were the epitome of naturalized treatments, insofar as naturalized treatments were those that addressed illness without recourse to magico-religious explanations. The opposite of such healing was not "artificial" pharmaceuticals but rather exorcisms, séances, incantations, sacrifices, prayers, and other such rituals, administered at temples to people who had been failed by the naturalized approach of physicians. (To further confuse issues, there was also the concept of *vis medicatrix naturae*, a Latin phrase for the Hippocratic observation that a sick person's body has its own natural ability to recover from harm and disease without excessive treatment or intervention.)

This secular, scientific naturalism clashes with the typical worldview of modern natural healing, which tends to be open-minded about supernatural and magical causes, welcoming a wide variety of spiritual approaches to illness, from shamanic healing rituals (offered at the Hippocrates Health Institute), to direct appeals to God, to the power of positive thinking. "Every human is a magician," writes don Miguel Ruiz in his massively successful bestseller *The Four Agreements*, touted as a book of ancient Native American wisdom and heartily endorsed by Deepak Chopra. "I see a friend and give him an opinion that just popped into my mind. I say, 'Hmmm! I see that kind of color in your face in people who are going to get cancer.' If he listens to the word, and if he agrees, he will have cancer in less than one year."[7]

Early proponents of naturalistic medicine would have rejected such a statement, whereas today it's most likely to encounter sympathy among natural medicine enthusiasts and censure from mainstream physicians.

The closer one looks at ancient medical practices, the harder it can be to see a clear connection to the specific protocols of modern natural healing. While the Hippocrates Institute promotes a raw vegetarian diet of "living foods," the Hippocratic corpus does not. Various meats were thought necessary to address patients' symptoms (puppy meat is "moistening" in case you're too "dry"), and cooked food was regarded as far superior to raw. In the distant past, reads the treatise *On Ancient Medicine*, people "suffered much and severely from strong and brutish diet, swallowing things which were raw, unmixed, and possessing great strength, [and] they became exposed to strong pains and diseases, and to early deaths." It was humans' improvement upon nature with food processing and cooking that allowed us to live full, healthy lives—in fact, the author of the treatise goes so far as to categorize cooking and food processing as a form of medicine.[8]

And what about the Hippocratic quote, "Let food be thy medicine and medicine be thy food," ubiquitous in natural healing publications and websites? Nonsense, according to Helen King, a classicist and leading scholar of Hippocrates. When I spoke with her, she expressed frustration with the tendency to uncritically ascribe one's favored position on medicine to Hippocrates, a practice that dates back for centuries. "He has been constantly reappropriated as whatever people want him to be," King explained. "Galen loved the humors, so he focused on the Hippocratic texts about humors. In the seventeenth century they believed climate caused illness, so they focused on that. There are so many different texts in the corpus that you can always pick out the one that you like." For King, Hippocrates's most popular quote represents the very worst version of this fallacious appeal to authority. "The text it supposedly comes from is almost impossible to understand," she told me. "It just says food-medicine-medicine-food, and no one knows exactly how to translate it. Certainly not 'let food be thy medicine and medicine be thy food.'"

Seeking consistency and historical accuracy in natural medicine yields only frustration, because natural medicine is not primarily a product of systematic thinking or historical scholarship. To throw up one's hands (as I once did) at the strangeness of calling acupuncture natural—"How is sticking stainless steel needles into someone natural?!"—is to miss the point of the word, and the corresponding ritual. The modern world of natural healing is the response of humans who seek totalizing explanations for their suffering; who seek integrity, physical and mental, after illness breaks them apart; who every day seek to affirm hope in order to live; and who do not have these needs met by the dominant medical paradigm. In short, the popularity of natural medicine is a reminder that it is *humans* who fall ill, not merely biological bodies, and that the standard healing rituals of purely scientific medicine can fail to recognize that fact. When seen in this light, the frustrating messiness of natural medicine, with its infrared saunas and wheatgrass shots and energy healing—well, it starts to make a little more sense.

In *The Illness Narratives*, the psychiatrist and anthropologist Arthur Kleinman argues that the experience of suffering forces two basic questions: Why me? (the question of bafflement), and What can be done? (the question of order and control).

To answer these questions, traditional medical systems invoke a model of the human being as a harmonious system, and illness as violation of that harmony. It's an intuitive model because it aligns well with the everyday experience of most people. Health is the default condition of the body, and illness represents a departure from a normal, balanced state. The resemblances between traditional systems in this regard are unmistakable. In Chinese medicine, health is thought to require a balance of yin and yang elements; in Greek and Roman humoral medicine, health depended on *eucrasia*, a balance of four bodily fluids.

Imbalance is a compelling explanation because it helps to answer the question "Why me?" The body's natural balance reflects participation in a natural cosmic balance, which is simultaneously physical and metaphysical. Physical and moral causality are thus linked up in a totalizing narrative. To be ill is to be in violation of the natural order. Illnesses of individual bodies mirror illnesses of the body politic and the natural world, since political disasters and natural disasters are also the consequence of imbalanced forces. Why me? Because you—or your ancestors, or your society—have violated the proper, natural order. The gods are being crossed; the ruler is corrupt; someone, somewhere, has upset the balance.

For most of history, people could find healers who provided these holistic, totalizing explanations of illness and health. In the nineteenth century, however, professional Western medicine underwent a radical shift. After the tremendous success of secularization in the biological and physical sciences, medical science began to follow suit, rejecting supernatural forces and moral disorder as causal agents. A watershed moment in this transition came in 1871, when Edward, the Prince of Wales, fell ill with cholera, and Queen Victoria asked clergy to pray for his health. Her once-uncontroversial request was met with scientific skepticism, culminating in British physicist John Tyndall's suggestion that the healing power of prayer be subject to experimental testing: one ward of a hospital with patients receiving prayers, and another as a control—then compare the mortality rates.

Tyndall had the narrow scientific objective of debunking the material efficacy of prayer. Yet, as documented by the historian James Opp, public debate at the time quickly turned into a broader discussion of prayer's role in people's lives. Was it really wise to undermine public confidence in the power of prayer? What was the cost of hardheaded empirical skepticism? Opp gives an example of a response by the Canadian writer Agnes Maule Machar, who stressed that humans require hope when their own knowledge and abilities fall short. For believers, argued Machar, prayer acts as ritual affirmation of trust in an "unseen Will whose love and care they have already seen revealed in the provisions of nature."[9]

Machar felt the narrow objective of Tyndall's experiment failed to recognize that the essential purpose of prayer goes beyond its material efficacy. Yes, people pray *for* certain outcomes, but paradoxically, achieving those outcomes isn't really what prayer is about. "Any mother's heart," declared Machar, would find it "infinitely more consoling to be told that the issue of the disease was under control of a wise and loving Father, who, though He acts in and through natural laws, has proclaimed Himself the Hearer and Answerer of prayer."

This was true, Machar maintained, even if the child were to die, for at least the mother could believe that God had "guided the event wisely." In other words, it doesn't matter if prayer works, in the strict sense, so much as it matters that the person praying can reaffirm the existence of a benevolent force at work in the natural world.

To complicate issues, the nineteenth-century crisis of faith over spiritual healing was mirrored in a crisis of faith over secular healing. Although scientists like Tyndall might have objected to Machar's emphasis on prayer, her references to the "provisions of nature" and "natural laws" would have met with more sympathy. At the same time that skeptics were debunking religious superstition, they were also debunking secular medical rituals such as bloodletting and mercurial emetics. "Doctors were killing their patients," wrote one physician in the early 1830s. "It would be better to trust to Nature than to the hazardous skills of the doctors."[10]

He was right. As the Harvard biochemist L. J. Henderson once remarked, it was not until the first decade of the twentieth century that "a random patient, with a random disease, consulting a doctor chosen at random had, for the first time in the history of mankind, a better than fifty-fifty chance of profiting from the encounter."[11]

Nature's healing power—the *vis medicatrix naturae*—became the preferred alternative to bleeding, vomiting, and prescription of alcoholic beverages. The recovery of patients, falsely attributed to these dangerous interventions, was now understood as the work of natural forces within the body. Writing in 1835, Jacob Bigelow, the president of the Massachusetts Medical Society, suggested that the medical community should put a "low estimate on the resources of art, when compared with those of nature."[12] According to Bigelow, it wasn't just faith healers and quacks who misled the public by taking credit for nature's power—doctors were also guilty. In most cases, he argued, the physician should be "the minister and servant of nature."

Physicians as ministers of nature, nature as benevolent healing force: it made scientific sense to understand nature as a stand-in for God, and natural healing as an extension of divine power, which we tamper with at our own peril. "A kind Providence has endowed every organ, and the totality of the organs comprising the system," with the ability to "return to a healthy structure and function," wrote a Massachusetts physician in 1863.[13]

Threatened by the encroachment of nature on their turf, some doctors accused Bigelow of attempting "to deify nature." But it was a losing battle, and in 1860 the great medical reformer and Harvard Medical School professor Oliver Wendell Holmes Sr. declared himself a firm believer in the "nature-trusting heresy."

Meanwhile, alternative healers appropriated this naturalized version of God and used it to offer people explanations and hope (or sell them quackery, depending on how you see it). In his 1925 book *The Healing Hand*, the "magnetic healer" Sidney Weltmer provides a perfect illustration of how traditional faith healing—laying on of hands—could be recast as science using the bridge concept of nature, which played the role of God while masquerading as a secular scientific idea. "The unconscious mind, or the healing mind—God's healing power in man—is the only healer of the ills of man," wrote Weltmer. "It is that force which has been called 'Nature' by physicians since the time of Hippocrates."[14]

Extending the associative logic, it made sense to encourage this force with "natural" medicines rather than pharmaceuticals. "Why use poisonous drugs," asked Joseph E. Meyer, an early twentieth-century botanical medicine salesman, "when nature in her wisdom and beneficence has provided, in her great vegetable laboratories—the fields and forests—relief for most of the ills of mankind?"[15]

Like ancient medical traditions, natural healing builds symbolic power out of parallels between the human body and the body of the world. At the time of medicine's secularization, the industrial revolution was roaring into full gear. Rivers flowed with man-made poisons—the arteries of nature's body infected by our artifice. As we devastated the natural world on an unprecedented scale, it became clear that industrialization was also damaging to human health. Smog choked the skies and our lungs; industrial pollutants sickened wildlife and humans alike; cities and factories doubled as petri dishes where new diseases flourished.

In this context it was reasonable to explain illness as a consequence of violating nature, since many illnesses really were caused by the same mechanisms responsible for destroying the natural world. Prominent figures such as John Harvey Kellogg and Sylvester Graham advocated for natural health by appealing to this obvious parallel. "God created man with a perfect constitutional adaptation to the state in which he first placed him," claimed Graham. "Man was constituted for the natural state, and not the artificial state, of civic life. [...] As an animal, man is constituted with the same physiological principles, as those which pertain to the constitutional nature of the horse, the ox, and other animals; and it is well known that these animals cannot be greatly diverted from their natural laws of constitution and relation without a deterioration of their natures."[16]

Though largely excommunicated from the texts of medical science, God and religious language remained in the writings of many early twentieth-century natural medicine advocates. Kellogg's 1903 book, *The Living Temple*, argued in the preface that "civilized man has departed far from the natural, the divine way of life." Graham frequently referenced "the supreme constitutional laws that God, in infinite wisdom and benevolence, has established in the nature of things." And in Germany, the naturopath Adolf Just managed to fit all these claims into the title his 1904 book, *Return to Nature!—The True Natural Method of Healing and Living and the True Salvation of the Soul—Paradise Regained*.

To practice the medical rituals of natural healing was, and still is, ritual affirmation of a unifying explanatory system. Then, as today, natural medicine answered the mystery of human suffering with a grand narrative and an ultimate cause. And then, as today, that narrative could transform otherwise dubious interventions into the tools of salvation. If you were an arthritis sufferer in 1907, you might use the popular Lambert Snyder Vibrator, as advertised in *Good Health*, a magazine edited for decades by Kellogg:

> When your head aches you rub your temples. Why? Because vibration is Nature's own remedy, and rubbing is Nature's crude way of creating vibration and starting the blood to going. Disease is only another name for congestion. When there is disease or pain there you will find the blood congested and stagnant.[17]

Physiologically, the ritual of using the Lambert Snyder Vibrator reduces to nothing more than a massage. But symbolically it does much more, explaining not merely the proximate cause of suffering but all suffering and disease. That so many people embraced and continue to embrace similar interventions underscores the rhetorical power of framing medical rituals as part of a totalizing explanation of illness. With natural healing interventions, the story you tell yourself is just as important as the intervention itself. It's a story that links up with a broader cultural narrative, weaving together a theory of bodily dysfunction with a theory of dysfunction in cultural and natural systems. As one Pennsylvania doctor asserted in 1860: "In the growth of Philadelphia and other large cities, much change has occurred, which tends to diminish the vital powers. [...] The air and water are never so pure as in good country situations; the food is not in so perfect a condition; to these must be added the sufferings of poverty, always greater in crowded populations."[18] By the early twentieth century this perspective was ubiquitous, as in this quote from Joseph Meyer's 1933 pamphlet advertising medicinal herbs: "To us it seems clear that, in the quest for the new and different or modern, much of real value has been sacrificed. . . . We have gone altogether too far and paid altogether too much."[19]

A century later, the rhetorical framework of natural healing has not changed. Parallels are still drawn between cultural failure to care for the natural world and physiological problems with own bodies. In the documentary about cancer that Brian Clement recommended to me, viewers are warned that "chemotherapy works like pesticides." Footage of sick humans is juxtaposed with footage of ailing ecosystems; packed cities are set side by side with packed slaughterhouses. It's no coincidence that the title of the Hippocrates Health Institute's official magazine is *Healing Our World*.

What has changed is the approach of modern scientific medicine, which now rejects all totalizing explanations of illness. Words such as "good," "evil," "God," "spirit," and "harmony" are no longer available to explain disorders of the body, just as they do not explain the emergence of mountains or the evolution of species. Proximate causes of illness remain, but the language of ultimate causes has been banished from professional medicine, as it has been banished from geology, biology, and physics.

As a result, the explanatory power of medical rituals, both diagnostic and therapeutic, is drastically circumscribed. "Why me?" is now a frequently unanswerable question, except in clear-cut cases such as lung cancer and smoking. Even in this case, a lifetime smoker might well ask why she ended up among the unfortunate 15 percent who get cancer, instead of the 85 percent who do not, to which any responsible oncologist can answer only one way: "I don't know." There is no longer a theory of inherent balance to which therapies restore the patient; no corresponding moral and spiritual balance in the world. Physicians have gone from being theologians and philosophers to scientists and mechanics.

The significant benefits of this shift should not be discounted. Like its cousins, medical science has progressed immensely because of its prohibition on supernatural answers to difficult questions. Our bodies really are complicated machines, and seeing them as such has opened up new avenues for research and treatment, resulting in extraordinary advances that include everything from organ transplants to mapping the human genome.

Not only that, a mechanistic approach to medicine is strongly exculpatory. In the last century we have discovered the biological bases for what were once understood as moral and spiritual failings. We no longer have to blame ourselves for our illnesses, or those of our loved ones, which can be an enormous relief. "Sometimes my patients will say, 'Is it because when I was forty-five, I did X?,'" Vinay Prasad, a hematologist-oncologist, told me. "In those moments, people have intense fear and regret, and when you say it's not that, it removes something they're really worried about."

However, Prasad's account, which I've heard from many physicians, does not merely illustrate the exoneration that patients might receive during the medical ritual of diagnosis. It also demonstrates patients' deep desire for an explanation. Although the exculpatory side of mechanistic medicine may sometimes come as a relief, for certain people the lack of an answer—even an answer that results in blaming oneself—can be traumatic. Paradoxically, the current "natural" healing movement is a rebellion against the explanatory deficiencies of fully *naturalized* medicine. Whereas the modern medical establishment generally offers only atomized diagnoses and therapies, natural medicine has a totalizing explanation built into its very name. Why me? Because you—or society—have departed from the natural order of things. What can be done? Return to the natural order. The two dominant questions raised by illness are answered according to a single unified theory, opening space for ritual control of the meaning of one's suffering.

The narrative of natural purity and unnatural corruption can lead to costly, dangerous rituals. Ineffective cancer treatments are only one among many examples. Equally worrisome is vaccine avoidance, which the sociologist Jennifer Reich has shown to be bound up with concerns over what is natural. Parents who worry about vaccines tend to be drawn to natural medicine more generally, as well as natural birth, natural eating, and other natural practices. The rejection of vaccines in favor of "natural living" and "natural immunization through breastfeeding" is about much more than physiology. According to Reich, parents who oppose vaccination see immunization as a ritual that affirms participation in a distasteful—unholy—way of life, so they eschew it in favor of what they see as more natural options. "Naturalness allows people to create an ideological schema that organizes information in moral ways," she told me. "Natural practices are a way of expressing anti-consumption and anti-corporate markers."

There is good reason to be infuriated by these practices. Here are people willing to sacrifice themselves, their children, and herd immunity on the altar of a false god, kooks and conspiracy theorists lining Brian Clement's pockets without anything to show for it. And the inconsistencies in their rituals are so flagrant: the supplements are processed; the cold laser is a *laser*, for God's sake!

And yet, after meeting with people at the Hippocrates Institute and listening to them in person, I found myself becoming more sympathetic. I, too, am frustrated by the thought of having to participate in a cultural system responsible for climate change, animal cruelty, and ecosystem devastation. I know that not long ago, industrial chemicals were barely regulated. I know that in the current era of deregulation and diminished enforcement, many still go unregulated. Millions are sickened by living near factories, industrial agriculture, and other emblems of unnatural modernity. I have read the same studies about super-bacteria created by our overreliance on antibiotics; about the danger of living in the constant glow of our screens instead of the rhythm of natural light. "You often hear, 'Oh, I grew up near this factory and swam in this stream,'" said Prasad of his cancer patients. "I don't just dismiss those ideas. We know what the EPA was like in the sixties, the rivers on fire." (Prasad isn't exaggerating: Cleveland's Cuyahoga River, polluted by industrial waste, caught fire for the first time in 1868, and did so thirteen more times, most recently in 1969.)

Confronted by systemic problems with existential consequences, it's no wonder that some people choose natural health rituals. Systems are immensely complex, and as humans we do not have the luxury of spending every minute analyzing how our actions fit into that complexity. We need a shorthand for figuring out how to live our lives. We need clear categories, stories, and corresponding rituals, especially when faced with a crisis like illness. Natural serves as shorthand for something like "outside of the dominant system," which explains the variety of indigenous, ancient, and non-Western rituals that are embraced under the label of natural healing. The category is not meant to demarcate a coherent set of rituals. Rather, it emerges organically out of basic human needs and anxieties, gathering together rituals that transport participants out of the dominant cultural narrative of healing, one that may be seen, not unreasonably, as neglecting the parts of ourselves that transcend biology.

In addition to ignoring the importance of explanatory narratives, modern medical rituals can also deprive people of their autonomy and dignity. These rituals often take place at a time when vulnerability and uncertainty are already breaking us apart, and by failing to actively affirm our humanity as an integral whole they exacerbate that breakdown. Once in the exam room you relinquish control over your body, following a script that under other circumstances would be unthinkable: disrobing in front of a complete stranger; allowing every part of your body to be touched and possibly penetrated, especially parts that are unusually sensitive or injured; disclosing intimate details of your life. Dutiful compliance with these rituals is expected from everyone—we become patients, which the *Oxford English Dictionary* defines as those who "endure pain, affliction, inconvenience, etc., calmly, without discontent or complaint, quietly expectant."

The journalist—and PhD in cell biology—Barbara Ehrenreich describes the dehumanizing experiences of being a patient as "rituals of humiliation." She recounts one of her own humiliating rituals, that of a pelvic examination by a male obstetrician, just weeks before she would give birth: "No words were exchanged until, when the speculum had been removed from my vagina, I inquired whether my cervix was beginning to dilate. He looked at the nurse and asked in an arch tone, 'Where did a nice girl like this learn to talk that way?'"[20]

Submission to a physician's authority is not only a process of physical humiliation. By visiting the doctor you are admitting, publicly, that you are incapable of taking care of yourself or your loved ones. Some people have no difficulty with this—after all, that's what experts are for. Visiting a mechanic is an admission that we cannot take care of our car, and it isn't (usually) a ritual of humiliation. But sick humans are not just broken machines. We have a deep emotional connection to our bodies. When your body breaks, it is *you* that is broken, not an object outside of you. This is why cancer patients routinely describe the experience of their illness in terms of being betrayed by their bodies. The physical symptoms of illness represent the collapse of an intimate relationship, a rupture of trust.

Yet precisely at the moment your body betrays you, medical rituals take it further from your control. When physicians used to handwrite prescriptions, their handwriting was famously inscrutable, a mystical formula meant only for the eyes of the pharmacist. The transition away from handwriting has done little to improve the underlying problem. Medicines function according to biological principles that the majority of patients tend not to understand. Even the names of medicines seem designed to be alien. If, like Ehrenreich, you are a breast cancer patient, suddenly you are thrust into a world where regaining control of your body depends on Abraxane, mitomycin, Taxotere, Xeloda, Thioplex, and Gemzar. Which of these is the best choice? Only your doctor can tell you, which means that choice really isn't involved at all. And why did your doctor choose Taxotere over Gemzar? Was it the right call? Now you must learn the arcane language of clinical trials, p-scores, and NNT versus NNH, and once you do, the randomized phase II clinical trial that justifies your doctor's choice might be hidden behind a paywall.

When the Canadian courts decided the hospital couldn't force J.J. to continue chemotherapy, her parents published a letter describing their feelings in a First Nations newspaper. The hospital wanted to "control" their "choice of health care," they wrote, which had far-reaching implications. By attempting to prevent them from choosing traditional medicine over chemotherapy, the "Haudenosaunee way of life was put on trial." Like the hospitals, the judicial system did not acknowledge their culture: "The courts also [proceeded] through our Big Green Corn ceremony, Karihwi:io and the death of an important figure in our society. It was disappointing that no one asked the courts to pause during these times because that is a societal norm for the Haudenosaunee."[21]

In his ruling against the hospital, Justice Gethin Edward emphasized that medical rituals constitute an integral part of First Nations identity. To justify his claim, he cited a paper submitted by the National Aboriginal Health Organization (NAHO), which includes the following creation story:

Soon after this new world had begun its transformation, the Sky Woman gave birth to a baby girl. The baby girl was special for she was destined to give birth to twins. The Sky Woman was heartbroken when her daughter died while giving birth to her twin boys.

The Sky Woman buried her daughter in the ground and planted in her grave the plants and leaves she clutched upon descending from the sky world. Not long after, over her daughter's head grew corn, bean, and squash. These were later known as the Three Sisters. From her heart grew the sacred tobacco, which is now used as an offering to send greetings to the Creator. At her feet grew the strawberry plants, along with other plants now used as medicines to cure illnesses. The earth itself was referred to as Our Mother by the Creator of Life, because their mother had become one with the earth.[22]

The myth shows how it is impossible to separate medical rituals from religious rituals. In the words of NAHO: "The crux of this story explains how the Haudenosaunee received their knowledge of traditional medicines—medicines that are used by the traditional healers in ceremonies and healings to this day. Traditional medicine, as practiced by Haudenosaunee people, is key to the health and survival of Haudenosaunee as a nation."[23]

Survival is not merely about biology. Medical rituals have significance beyond clinical efficacy, functioning to reaffirm one's personal identity, cultural identity, and most deeply held beliefs. As Mohawk scholar Christopher Jocks argues, "Traditional ceremonies and spiritual practices . . . are precious gifts given to Indian people by the Creator." Such practices—among which medical rituals count as some of the most important—"afford American Indian people of all [nations] the strength and vitality we need in the struggle we face every day."[24]

Reconfiguring the meaning of medical rituals in these terms helps to explain the attractiveness of HHI to Makayla and J.J.'s parents. The epistemology of modern medicine does not accord any special value to traditional medical rituals, but natural medicine does. There is very little critique or exclusion in the world of natural healing. Ayurveda and traditional Chinese medicine, Haudenosaunee healing and homeopathy—despite being developed according to radically different principles—all are welcome because all are sacred to the people who believe in them. Calling them natural simply lays a symbolic and epistemological foundation for their coexistence.

Observing that modern medical science cannot do this is not a critique, so much as recognition of its necessary shortcomings. Methodologically separating the cultural and personal significance of medical rituals from evaluations of their efficacy is an extraordinary engine of discovery. But it also means that people who seek to justify medical rituals on different grounds will feel demeaned, neglected, and disempowered.

At HHI, by contrast, the focus is squarely on empowerment. "We don't say there's one absolute way to heal," Clement told me in our interview. "And we don't heal anyone. They heal themselves." These principles are built into virtually every ritual endorsed by the institute. From wheatgrass juicing to infrared saunas, one of the key distinguishing elements of natural healing is that it gives people the ability to heal themselves. According to the principles of natural healing, you can own—both metaphorically and literally—the means of healing illness. You do not need medicines that are available only by prescription, or machines that can only be purchased by hospitals, or knowledge that is only available in medical school.

The rhetoric and rituals of natural healing transform patients into active agents. As Jennifer Reich observes, "vaccine resistance represents parents' insistence they are experts on their children and empowered to challenge professional recommendations." The same is true for "natural" supplements and dietary regimens, which two professors at Harvard's School of Public Health describe as "daily activities of affirmation and assurance . . . liturgical acts of recognition with deeper implications for social, moral, and spiritual redemption."[25] In a sociological study of wellness culture, the scholar of rhetoric Colleen Derkatch confirms that for most of her subjects, seeking out natural medicine is intimately connected to their need for "empowerment."[26]

As someone who has never faced a life-threatening disease or a chronic illness, whose loved ones are mercifully healthy, and whose beliefs about medicine enjoy cultural dominance, it is easier for me to discount the need for empowerment. I shouldn't. Disempowerment is a glaring flaw in the standard modern approach to healing, as evidenced in the testimony of patients who seek autonomy and dignity through other medical rituals. In that sense, Makayla Sault's death at Hippocrates can be seen, at least in small part, as a symptom of disorder in the dominant system. Preventing others like it will not be accomplished by shutting down the Hippocrates Institutes of the world—not, at least, without accompanying reforms of the standard approach that drives people to seek them out.

After interviewing Brian Clement, I was given an informal tour of the premises by a naturopath. As we walked from the wheatgrass juicing room to the greenhouse, she told me about multiple cancer patients whom she'd personally seen recover using natural therapies after being told there was no hope. "The doctors sent them to hospice," she said, "so they came here, and they were healed naturally. When they went back, the doctors said it was like a miracle."

Everyone at Hippocrates has a miracle story. Sometimes the stories are personal: complete recovery from psoriasis that failed to respond to "artificial treatment with pills and steroids." Other stories are secondhand: a cousin cured from stage IV prostate cancer, a friend no longer taking any medications. I soon discovered that sharing tales of hope was a central ritual of the community. The *Healing Our World* magazine regularly runs profiles of people who apparently triumphed against all odds. "Enjoy 13 inspiring nitively explaining suffering belongs to theologians, not physicians. en-mindedness about people's sacred rituals may be a virtue, but doc-s and scientists should repudiate anecdotes and practices that are not ported by evidence, religious or otherwise. Hope is important, but it's rly malpractice to promise, as Clement does, that thanks to the healing ver of nature, "everything you struggle with can be easily solved."

However, learning from natural healing does not mean peddling snake or declaring faith in the transcendent power of nature to cure everyone's nents. Ultimately, what natural healing really provides is a diverse set of als that treat illness as an existential crisis, not merely physiological dys-ction—rituals that address the full humanity of sick individuals instead of ucing them to patients. This can be a good thing. There are differences ween HHI and regular hospitals that do not reflect superstition. Take, for tance, the ubiquitous presence of plants. To be at HHI, whether indoors outdoors, is to feel connected with nature, surrounded by living greenery, a time when one's connection to life feels especially tenuous. Hospitals, contrast, are uniquely devoid of life. With few exceptions they are sterile, less mazes of blinding white and stainless steel, with not a plant in sight—ally not even a painting of one.

I asked Prasad about the absence of plants in hospitals. Was there a entific basis for it? "In some exceptional situations, like a bone marrow nsplant, I can understand banning plants," he said. "But in the majority of dical settings I think it would be fine." Even supposing that plants are un-eptably dangerous presences in doctor's offices and waiting rooms (some ients are quite sensitive to scent, pollen, and other natural allergens), there ld certainly be more atriums and hospital gardens. Multiple studies have wn plants to be beneficial for patient health and well-being. Healthcare hitecture specialist Roger Ulrich has documented how gardens and plants re common in hospitals until the 1900s, when an emphasis on "functional iciency" turned hospitals into their modern incarnation: "starkly institu-nal, unacceptably stressful, and unsuited to the emotional needs of patients, ir families, and even healthcare staff."[29] An entirely unsuperstitious take natural healing would recognize the importance of being around life—of ilitating hospital garden walks, say—instead of systematically excluding it.

Cultural practices can also be welcomed into hospitals without endan-ring patients or the integrity of scientific medicine. Consider an inner-y primary care clinic in Vancouver, Canada, where researchers decided

stories of miraculous wellness," advertises the cover of one
readers learn about the salvation of a woman named Alice.
decades ago, [...] the top hospitals for cancer in the U.S.
Sloan Kettering, Moffitt) told Alice that her demise was er
amputated her leg—but even that would only add six mo
Instead of heeding the hospital's advice, Alice's "loving hu
to Hippocrates to participate in the Life Transformation Pi
"By the end of their stay, Alice was no longer in a wheelcha
ing with [her husband] as the Wheat Grassband played." 1
later, both members of the couple are flourishing and ha
viewed about their story.[27]

Seeking hope in the face of illness and death is anothe
it means to be human. Stories like the ones in *Healing Ou*
scribed on stones outside of healing temples in ancient G
sick could read about those who had visited physicians with
to be cured by priests. Catholic saints are said to perform l
and many Buddhists have long believed in the healing powe
pearls recovered from the ashes of cremated Buddhist maste
week poll found that 72 percent of Americans believe "pra
cure someone—even if science says the person doesn't stan

Alongside empowerment and explanation, the appeal
ing can be partially located in its hopeful affirmation of a
vine force, present everywhere, that seeks to promote life ai
health. Ritually affirming the existence and power of this fu
what bioethicist Insoo Hyun calls "spiritual distress." Severe
ness can isolate you from the sources of meaning in your lif
live in a different world of concerns from your friends, con
ity and suffering in a way that many around you are not. In
writes Hyun, the antidote is "therapeutic hope" in the form
that, in addition to promising efficacy, also create meaning
in crisis. Nor is spiritual distress exclusive to the severely
experience a mild case of it at the thought of potentially con
because of exposure to environmental pollutants. In that c
peutic hope in the form of an herbal supplement could treat
brief prayer to nature that you may be kept from harm.

It might seem as if modern medicine has nothing to
messy, unsystematic, quasi-religious world of natural heali

to study the impact of including indigenous Elders in the treatment of First Nations patients. The results were striking. In addition to clinically and statistically significant reductions in depressive symptoms and suicidality, patients provided first-person testimony about how having Elders present transformed their experience of medical care: "Just being able to sit down with somebody that recognizes you as a human being and also has the cultural education of . . . sage and cedar and the natural medicines that biologically [are] part of our universe has made a huge difference for me."[30]

One of the most commonly reported effects of having Elders present was an improvement in patients' trust—another area where modern medicine is falling short. In a remarkable decline, trust in medical leaders has gone from over 75 percent in 1966 to only 34 percent today. Unfortunately, a big part of the problem is that good reasons exist to distrust the medical establishment. "People are right to say that the trial agenda is perversely driven by financial interest," Prasad told me. His book *Ending Medical Reversals* exposes how merely promising medications, as yet unproven in rigorous clinical trials, often receive early approval, only to be shown ineffective. And yet, when severe problems with medications are exposed—think the opioid crisis—the people behind them get off with a slap on the wrist (if anything), which advocates of natural healing are quick to point out constitutes a double standard underwritten by money. "I trust natural medicines more because companies can't patent them," one patient at HHI said. A rigorous, systematic, and highly publicized effort to crack down on the problems that Prasad identifies would go a long way toward earning back the public's trust, rendering people less vulnerable to individuals like Clement. Such efforts are already underway in the areas of over-screening (routine mammograms are no longer endorsed for all women) and over-prescription (antibiotics are no longer routinely prescribed for infection). The highly esteemed *British Medical Journal* even has an initiative called "Too Much Medicine," which unites researchers, clinicians, and policymakers who want to fight against "uncritical adoption of population screening, disease mongering and medicalisation, commercial vested interest, [and] strongly held clinical beliefs," all of which fuel public mistrust.[31]

Of course, Elders themselves don't earn trust by cracking down on ineffective medicines and overmedicalization—they earn trust by speaking in a language that indigenous patients find familiar and empowering. Healthcare professionals can do the same thing, recognizing that for many people, framing medical interventions as natural can be empowering. Reich interviewed one pediatrician who described using the word "natural" to reassure parents: "I say, 'You're introducing a very mild . . . form of this illness to your body, allowing your body's natural immune response to create protections for you for the next time you encounter this virus, or this illness. So your body is producing its natural reaction.'"[32]

Another way to use nature as reassurance is by recognizing that death itself is natural, as Barbara Ehrenreich did when confronted with her cancer diagnosis:

> Somehow, when I listened to the song of a blackbird in the garden, I found it incredibly calming. It seemed to allay that fear that everything was going to disappear, to be lost forever, because I thought, 'Well, there will be other blackbirds. Their songs will be pretty similar and it will all be fine.' And in the same way, there were other people before me with my diagnosis. Other people will have died in the same way I will die. And it's natural. It's a natural progression. Cancer is part of nature too, and that is something I have to accept, and learn to live and die with.[33]

No set of institutional changes will fully satisfy every person who experiences spiritual distress. Medical ethics dictate that therapeutic hope cannot veer into false hope, which means medical science should remain institutionally constrained by honesty about the limits of its power, in ways that other healers are not. Divine healing forces, even if they are disguised as nature, are the exclusive province of priests, not physicians. But rejecting the religious side of natural healing does not entail rejecting all of it. Neither magic nor supernatural power is required to transform rituals of humiliation into rituals of empowerment. All it takes is a little more attention to human nature.

CHAPTER 6

Deepak Chopra's Condo

WHILE WAITING FOR A SALES AGENT at Muse Residences in Sunny Isles Beach, Florida, I picked up a prominently displayed copy of Deepak Chopra's latest book. It was the physician and self-help guru's 2017 instant bestseller, *You Are the Universe: Discovering Your Cosmic Self and Why It Matters*—unread, and a good thing, too. Otherwise, prospective condo buyers might have come across an unfortunate metaphor on page 171, meant to illuminate the structure of hemoglobin. The hemoglobin molecule, we learn, is extravagantly complex, a gigantic shell that exists for the sake of only a few key atoms. "Think of rich people living in huge mansions," writes Chopra, "that rationally speaking are a waste of space for one or two people to rattle around in."

At the Muse office it's difficult not to think of rich people and their extravagant structures, since that's precisely what's on offer. Towering above the oceanfront, the gleaming fifty-one-story ultra-luxury condo building boasts "standard" amenities like a private fitness center and spa, along with a private movie theater, a wine tasting room, and a fully automated parking system that takes up the first few floors. According to the agent I spoke with, a significant number of buyers will be using the condos—which start at just under $5 million and top out at $17.5 million for a six-thousand-square-foot penthouse suite—as vacation homes.

If you've got another $500,000 to spare, you can upgrade your Muse condo to "wellness residence," and that's where Chopra comes in. He's been working with the design company Delos on the details of this upgrade, advertised as a suite of "scientifically validated wellness solutions" that "naturally support our body systems and rhythms." Among them: advanced water purification; dynamic lighting for circadian alignment; whole-home IAQ sensor platforms to detect air pollutants; and even "hand-picked Chopra finishing selections including mood aligning paint colors mimicking nature."

The beautifully produced catalogue assures you these interventions will be profoundly transformative. Wellness Real Estate™ has done nothing less than "completely reinvent how we should all be living." Securing Chopra's involvement—he himself owns a $14.5 million Delos condo in New York City—is meant to demonstrate that $500,000 buys you an upgrade in mind, body, and spirit. A Delos home makes everyday life sacred by aligning you with nature's intended rhythms. Every breath, every vitamin-C-and-aloe-infused shower becomes a purification ritual. As Robin Finn put it in the *New York Times*, the pitch is for "an empathic multimillion-dollar home that passively treats the occupant's body like a temple."[1] A temple that treats you like a temple.

The flipside of the promise—*completely reinvent how we should all be living*—is that most people are *not* living as they should. The majority of us are out of sync, unwell. "The fundamental elements of our lives—air, water, light, and comfort—are often underappreciated until experienced in their purest state," reads the catalogue copy. By accepting that you can purchase this purity, you also accept that public air and public water are inferior, lesser versions of their ideal forms. To be truly blessed, the residents of Delos homes require the existence of their cursed opposites. These opposites live in a polluted world, both corrupt and corrupting. The curated solutions at Delos offer an escape from that corruption by transforming domestic life into a purifying religious practice aligned with nature. "We can restore ourselves at home," says the catalogue. Restore ourselves from what? The undesirable grimy normalcy to which everyone else is condemned. The author of *You Are the Universe* is there to remind us of the spiritual stakes, uniting the material and the metaphysical in an endless feedback loop of natural goodness. Chopra lives in a natural Delos home because he is a wise sage; he is a wise sage because he lives in a Delos home that aligns him with nature.

When natural living is a luxury commodity, the act of making money, lots and lots of money, turns into a sacrament. The wellness upgrade doesn't just transform your shower into a religious ritual; it transforms the work you did that paid for the shower. The ultra-rich enjoying their Muse vacation residences can feel secure in the belief that accumulating great wealth is not mere hedonism, but rather an exclusive means to holiness.

Gwyneth Paltrow's lifestyle brand, Goop, performs the same alchemical conversion of wealth into natural holiness, best illustrated by Goop's first fragrance, "Church." Like all Goop fragrances, Church is "composed entirely of rare, all-natural elements imbued with the power to entrance, heal, and transform."[2] Natural sourcing and production methods take "niche artisanal perfume to a new level" by granting customers "the mystical, homeopathic, and Ayurvedic benefits of pure plant essences." (Chopra centers his mystical life philosophy and healing practices in Ayurvedic understandings of nature.)

Whereas other scents are harmful products of the "synthetic chemical industry," Goop fragrances are "pure, alive—the real thing." The real thing doesn't come cheap. Church perfume runs $165 for a 1.7 oz. bottle (Chanel No. 5 is a comparative bargain at $105), and a Church-scented candle costs a whopping $72. Speaking to students at Harvard Business School, Paltrow made it clear that the high price is essential to the brand. When pressed about elitism and the importance of engaging with lower-income demographics, she remained firm. "It's crucial to me that we remain aspirational," she said. "Our stuff is beautiful. The ingredients are beautiful. You can't get that at a lower price point. You can't make these things mass-market."[3] A few moments on Goop's site and you quickly understand it isn't just beauty, or even primarily beauty, that's set up as aspirational. It's also natural purity and holiness. Being real becomes a luxury.

Like Delos's Wellness Real Estate™, Goop sets up a contrast between what it's selling and the unclean, unnatural status quo. Its mascara is "clean, natural, and organic," unlike conventional mascara. In fact, the premise of Goop's entire "clean, non-toxic" line of beauty products—including the Phyto-Pigments Ultra-Natural Mascara—is that conventional products can't be trusted. They are toxic and tainted, and they pollute the people who use them. Even if you try to buy "natural" products, you'll fail in a conventional store, since the people running conventional stores are likely as tainted as the products they sell. "Because it's a free-for-all," cautions one Goop blog post, "companies can use whatever adjectives they'd like when it comes to marketing and 'greenwashing'—*natural, green, eco* literally have no enforceable definition."[4] The "average" American woman ends up inhaling, ingesting, and absorbing more than one hundred toxic chemicals through personal care products, even when she has tried to purchase safe ones. Only

Goop can protect you. Purchasing their products, no less than using them, is itself a purifying ritual, a connection to honesty and truth in a world of deceivers. When you buy your mascara, your good feeling depends on a tacit contrast to other shoppers who are impure, with minds, bodies, and spirits made unclean by the unnatural products sold everywhere else.

As we already have seen with food, when "natural" is a theological term, buying natural must mean you are doing good in every possible way. Choosing to follow God's will is never wrong. Naturalness unifies all virtues: what is good for your skin must also be good for the planet. This is reflected in the sense of mission and altruism that's foregrounded throughout Goop's articles. "Those of you who have read *goop* for a long time know that we try to do well by our bodies, our kids, and the environment as much as possible," begins another blog post about clean, nontoxic beauty. Shopping elsewhere hurts you and the environment. By contrast, Goop's natural products benefit the purchaser and the world. For $58, you can buy 5 oz. of Naturopathica Oat Cleansing Facial Polish. The product cleanses your face while the purchase cleanses your soul thanks to a "major plus" built into the ethics of every bottle: "The apothecary-chic packaging is made of recycled materials whenever possible—the tubes are made from 70 percent post-consumer recycled plastic, while the cartons are made entirely from post-consumer waste recycled fibers." Pampering yourself is really altruism!

Delos does the same with its Wellness Real Estate™ by emphasizing the altruistic mission of its advisory board. Among the celebrity members is Leonardo DiCaprio (who owned a condo and then sold it), "well known for his dedication to the environment on a global scale," whose affiliations also include the "World Wildlife Fund, Natural Resources Defense Council, and International Fund for Animal Welfare." To participate in Delos is to be a member of a larger movement focused on preserving the natural world. If you care about being natural, you care about nature. Purifying your home by aligning it with nature is a ritualized version of purifying the natural world.

There is no more successful example of strategically linking the purchase of expensive natural products with ethical purity than Whole Foods Market. At the entrance of my local store, soaps regularly feature as the impulse buy of the day, neatly stacked on wooden crates and burlap, framed by advertising copy: "Fair Trade." "Give Everything Good." "Whole Planet." Normal soaps are usually emblazoned with their name—*Dove* or *Irish Spring*—but these soaps, made by Alaffia, swap the brand name for "good," pressed in large friendly letters on each bar and meant in the broadest way possible. "Alaffia balances science with tradition to holistically benefit your body, communities, and the environment," explains its website. Profits help fund "empowerment projects" focused on the West African nation of Togo, ranging from a maternal health initiative to school construction. Its "good" brand soap (biodegradable and gluten-free) is available exclusively at Whole Foods, and 100 percent of the proceeds support families and communities around the world. On sale days, if you're an Amazon Prime member you can purchase this goodness and cleanse yourself with it for just one dollar.

The soaps are part of a wider set of products that qualify for Whole Foods' Premium Body Care™ seal of approval.[5] The store's official description of the seal describes how premium body care standards were developed as an alternative to the word "natural" as it is used by other retailers. Customers are looking for "safer, more natural body care products," but "there is a lot of confusion about what 'natural' actually means." At regular supermarkets, personal care products labeled "natural" may contain harsh preservatives or "ingredients with environmental concerns." There's simply no way to know. But at Whole Foods, the standards are "more stringent than any other national supermarket." Premium body care products are guaranteed to meet the strictest standards for quality sourcing, environmental impact, results, and safety. It's a natural label that you can trust. Like Goop and Delos, the natural goodness of Whole Foods is predicated on the unnatural badness of the alternative.

This is *consecrated consumption*, in which the ritual of shopping becomes a kind of spiritualized retail therapy dedicated to nature. Whole Foods sanctifies my shopping experience by uniting claims about material quality and ethical quality under the rubric of what's natural. Natural personal care products are easy on your skin and the environment. Unlike other supermarkets, your fish will never be artificially colored, and you can choose it based on a sliding sustainability scale. Meat is labeled according to animal welfare. This is a place with values; the people who shop there share a mission. I know the shopping experience feels nice because I participate in it regularly. At Whole Foods, every purchase, even if it's unnecessary, is nevertheless a good one.

As someone who engages in the consecrated consumption of natural goods, I am intimately familiar with the anxieties that drive it. Many of these anxieties are selfless, starting with a love of the natural world and concern

for its well-being. I'm disgusted by the existence of the Great Pacific garbage patch, and frustrated by the thought that microplastics from products
I've purchased might be a part of it. I know unsustainable agricultural practices are destroying the environment, and I hope to distance myself from
them. These selfless concerns connect to self-interested ones. I know that
degradation of the natural world can have devastating consequences for our
health. When I lived in China, my daily jog left me with gray mucous and
a hacking cough so bad that I was forced to give up on exercising outdoors.
Couldn't the same pollutants that sickened me be present, albeit in lower
quantities, in the air I breathe as I write these words? Isn't it possible they are
bad for me? And what about water? Can I trust what comes out of my tap?
I do, but after the water crisis in Flint, Michigan, I understand why some
people wouldn't. How one cares for oneself is a mirror image of how one
cares for the world, and in both cases a wanton disregard for nature seems
to be a central problem.

Since corporate disregard for the natural world is clearly related to cost-
cutting, there's an intuitive logic to the high price of natural goodness. Companies refuse to spend money on better—that is, better *morally*—production
methods, opting instead to desecrate nature. Rather than clean up toxic
waste, just dump it in a river. Rather than use renewable natural resources
for food dye and makeup, just synthesize cheaper pigments from petroleum.
We who, understandably, do not want to be like the cost-cutting companies,
we need to pay up, even if they refuse to. Cleanliness and purity require
sacrifice. Those of us with the means can make that sacrifice: higher prices
for better homes, better products, a better life, a better world. Whole Foods'
hashtag is #MakesMeWhole; their nonprofit, to which my soap purchases
contribute, is #WholePlanet. As for those who can't make the sacrifice? It
may not be their fault, but that doesn't change the fact of their complicity
in a destructive system, and the resulting moral impurity of their unnatural lives.

Add to this the depressing truth that money really can buy the absence
of pollution. Property values are lower near factories, refineries, and highly
trafficked roads, which means lower-income people are more likely to be
breathing dangerously high concentrations of particulate matter. When water in a low-income community is contaminated with fertilizer runoff or
hazardous waste, it's difficult to force action on the problem because legal
help and expert research are expensive, not to mention time-consuming.

Extreme poverty can lead to pollution—so it makes sense, on the face of it, that extreme wealth would lead to purity.

Yet the long and dubious history of associating religious purity with socioeconomic class should give us pause when it comes to the latest naturalized version of the same. The conversion of class structure and wealth into spiritual status usually involves a corruption of the values that supposedly animate a religion. When class and purity are interchangeable, a religious system cannot offer solutions to injustice. Instead, it offers an explanation: a natural hierarchy in which the highest are necessarily pure and the lowest are not.

One of the world's oldest religious texts is the Rigveda, a collection of sacred hymns central to Hinduism. Hymn 10.90 narrates a famous creation myth, in which the gods sacrifice a divine giant named Purusha. Dismembered, Purusha's body parts become different elements of the natural world: "The moon was born from the mind, from the eye the sun was born; [...] from the head the heaven issued forth; from the two feet was born the earth."

In this same myth, Purusha's body parts also become different social groups. "When they divided Purusha, in how many different portions did they arrange him? What became of his mouth, what of his two arms? What were his two thighs and his two feet called? His mouth became the brahmin; his two arms were made into the *rajanya*; his two thighs the *vaishyas*; from his two feet the *shudra* was born." The social world reflects the natural world; the organization of humans into four groups—*varnas*—is a natural order.

Eventually the *varna* system was intertwined with *jati*, or the caste system, which organized humans into a spiritual hierarchy of purity. While the concepts of *varna* and *jati* in Hinduism (and beyond) are historically complicated, and the historical origin of Hymn 10.90 remains contested, there's no doubt that a hierarchal and religiously justified understanding of purity was commonplace in India (and to a certain extent still is). This hierarchy functioned to support the socioeconomic status quo that included not only class but also gender, with women being more connected to pollution than men, and therefore relegated to a lower status.

Ritual practices emerge from the purity hierarchy and help to reinforce it. In areas where caste is emphasized, lower castes cannot eat the same foods as upper castes. They do not wear the same clothes or work the same jobs.

Sometimes lower castes cannot even occupy the same space or touch the same object as someone of a higher caste without polluting them. Menstruating women cannot enter temples and they cannot cook. Endogamy, or the practice of marriage within a specific group, ritually preserves this hierarchy across successive generations. The taint of impurity threatens bloodlines as much as it does individuals.

When it is a function of class, religious purity becomes expensive. In ancient times processing food was a costly endeavor, so refined foods represented the pinnacle of purity, destined for the palates of refined people. The clarification of butter into ghee is a frequent metaphor for the purification of the spirit. "Since diet largely determined moral and intellectual standing, it followed that the humble were thought to have little chance to be virtuous," writes the historian Rachel Laudan in *Cuisine and Empire*. "For that, the cuisine of the higher ranks was a requisite. Eating refined, well-cooked food was thought to make one strong, vigorous, beautiful, intelligent, and thus virtuous."[6] These foods included pure white sugar, pure white rice, and pure white bread; and everything had to be cooked, even fruits—cooking fuel being a significant expense for all but the wealthy. To the cost of pure food may be added the cost of keeping domestic spaces pure; the cost of cleaning oneself and one's garments; the cost of sacrificial ceremonies (in which ghee might be used), all of which speak to the necessary relationship between holiness and money, ambiguously underwritten by a natural order.

An anthropologist could hardly invent a better modern analogue of this general understanding of purity than Goop. Instead of a Brahmin laying out a self-serving definition of purity, a celebrity laying out a self-serving definition of "clean." Consuming clean products, from food to makeup, grants the same qualities that eating refined food once did: strength, vigor, beauty, intelligence, and thus "wellness"—a seemingly secular term that avoids the judgmental connotations of virtue and holiness, but means essentially the same thing.

As Carl Cederström and André Spicer argue in *The Wellness Syndrome*, wellness is no less than "the moral demand to be happy and healthy." The demands of wellness, they observe, have a clear subtext rooted in a religious understanding of natural living. If you don't seek to "get back to some kind of imagined original state (whether that be Paleolithic life, biblical times or the Garden of Eden), then you are a morally defiled person."[7]

Goop's "Church"-scented candle shows how shifting definitions of defilement and purity are a function of social class. A couple thousand years ago, spices, perfumed oils, and incense were extremely costly. Their use in rituals was simultaneously religious and an expression of economic prestige, as in this passage from the Book of Esther: "Before a young woman's turn came to go in to King Xerxes, she had to complete twelve months of beauty treatments prescribed for the women, six months with oil of myrrh and six with perfumes and cosmetics." Back then, burning the Church candle would have been a sign of wealth simply because it smelled nice. Scent remained a sign of wealth and purity until well into the twentieth century. Writing about "the real secret of class distinctions," George Orwell summed up the difference between bourgeois and working class in four words that he heard regularly in his youth: "The lower classes smell."[8]

Smelling good is no longer expensive, though. A Yankee brand candle twice the size of Goop's can be had for a quarter the price. Consequently, the important distinction between upper and lower class no longer lies in the smell of the Church candle but rather in its naturalness, which is precisely where Goop locates its power to purify you, physically and spiritually—and also, uncoincidentally, what justifies the higher price tag. The same is true for all of Goop's offerings. The obstetrician and author Jennifer Gunter described the In Goop Health symposium as a series of pseudoscientific workshops dedicated to the purifying worship of "Goture," Gunter's portmanteau of God and Nature.[9] Tickets for the one-day event start at $1,000.

Avoiding defilement requires adopting a lifestyle that is, to use Paltrow's term, aspirational. That's why changes in the definition of clean eating—and, likewise, the culinary sources of ritual purity and defilement—can usually be indexed to changes in price. Whereas lower classes once ate cheaper *un*refined food, their modern analogues are now more likely to eat cheaper refined food. In a Goop blog post, "super-clean, almost monastic meals" are contrasted with a polluting meal that might be more recognizable to your average non-monastic American: "grilled cheese and fries." Whereas lower classes once drank water sourced directly from nature, the cheapest water now comes from taps. Hence the strange phenomenon of "raw water," where people pay a premium for exclusive bottled water that's "pure" in virtue of having never been processed, at least if you exclude the bottling and shipping process.

Spirituality is always part of the sell. A typical "raw water" company, Tourmaline Spring of Maine, touts its "sacred living water" as "verified naturally pure by science" and "filtered by mother nature to a degree that no man-made machine is capable of replicating."[10] Alongside the appeals to nature are appeals to exclusivity, starting with the bottle's blue ombré label that features a watermelon tourmaline. The water is not merely natural but "flawless, gem-grade spring water" that "is bubbling up of its own accord through gemstone-lined vaults in one of the most ancient mountain ranges of North America. The taste profile and sensation it creates in your being will prove to you how rare this substance is." So, for that matter, will the price: a twelve-pack of one-liter Tourmaline Spring bottles will set you back sixty dollars with shipping from Maine, which works out to nearly five dollars a bottle. Compare the "naturally purest spring water on the face of the planet" with San Pellegrino sparkling water, which can be purchased for around two dollars a bottle at most supermarkets, or to regular tap water, which is almost free and available to anyone with a tap: only spiritual fulfillment could justify the difference in price.

And if you should happen to eat a grilled-cheese sandwich or wander outside your Delos condo into the impure world? No worries, at least if you have funds. Purification is possible but expensive: A ten-day detox supplement kit from Goop is $169; a six-day detoxifying "perfect health" retreat at the Chopra Center, a mere $3,795.[11] The poor, like lower castes, are out of luck. They eat polluted foods every day, live in polluted spaces, and almost certainly cannot afford to detoxify themselves.

From a modern perspective it's tempting to explain all purity rituals, even ancient religious ones, with appeals to scientific principles. On this view, Moses's dietary rules exclude animals likely to have parasites. Washing as a form of spiritual purification exists because it helps to prevent the transmission of disease. Such explanations are now widely recognized by scholars as deficient. In her celebrated study *Purity and Danger*, the anthropologist of religion Mary Douglas systematically dismantles the "medical materialism" approach to ritual, cautioning that even if there are hygienic benefits to religious dietary laws, it is "a pity to treat [Moses] as an enlightened public health administrator, rather than as a spiritual leader."[12] Many purification rituals—exorcism, sacrifice to a deity, sprinkling with holy water—have no obvious hygienic explanation. Other rituals, such as thoroughly purifying oneself after merely touching a menstruating woman, combine hygienic intuitions (blood transmits illness) with socioreligious norms (women are lesser) to yield a result that cannot be fully explained by hygiene.

Douglas observes that we are most tempted by medical materialist explanations of secular purity rituals. Paltrow's ultra-natural mascara must be a response to the dangers of unnatural chemicals; Chopra's condo, with its filtered air and water, must be a response to the polluted world we live in. But modern secular purity rituals, like ancient religious ones, are only partially explained by hygiene. Chopra's condos celebrate their owners' unique power over nature. They focus exclusively on the commodification and control of nature, not submission or integration with it. That's why the wellness upgrade includes a lighting system and a thermostat instead of a rooftop campsite for sleeping under the stars. Similarly, Goop connects clients to nature through their ability to purchase it. Rather than embracing the natural reality of aging, they spend $90 on 3.8 oz. of an all-natural anti-aging "miracle oil." When natural holiness is a business, you ritually honor Mother Nature by buying things with her name on the label.

Another problem with linking wealth and spiritual purity is that possessing wealth can come to be seen as the necessary outcome of good living. The prosperity gospel movement, with its forty-thousand member megachurches and wildly wealthy pastors, preaches that prayer and piety will lead to riches. Flying in a private jet isn't an extravagance; it's evidence of election. The same principle applies for Chopra, who deftly naturalizes prosperity theology. In *The Seven Spiritual Laws of Success*, Chopra promises readers a path to perfect health and unlimited wealth "based on natural laws which govern all of creation." Does your socioeconomic status matter to your ability to be healthy and wealthy? Not if you live a morally pure life, here defined as natural: "Once we understand our true nature and learn to live in harmony with natural law, a sense of well-being, good health, fulfilling relationships, energy and enthusiasm for life, and material abundance will spring forth easily and effortlessly."[13] If you're really living a pure life in accordance with natural laws, it'll show in the size your bank account and the glow of your cheeks.

But when you subscribe to Chopra's email list it's immediately apparent that his substantial wealth also depends on the more mundane laws of marketing. I received endless hard-sell emails for expensive courses and even more expensive retreats. "Do you think you're worth investing in?" asks the subject line of one typical email, an advertisement for his SynchroDestiny course ("only six payments of $75"). "[SynchroDestiny] is the only way to truly achieve the most you want out of life." The method and tone might as well be copied directly from prosperity gospel icon Joel Osteen (net worth roughly $40 million), who exhorts readers to pay up or be consigned to spiritual and, thus, fiscal and physical, misery. "You can't afford to *not* tithe," warns Osteen. "If you will dare to take a step of faith and start honoring God in your finances, He'll start increasing your supply in supernatural ways. . . . He'll cause you to get the best deals in life. Sometimes, He'll keep you from sickness, accidents, and harm that might cause other unnecessary expenses."[14]

The history of religion is replete with warnings about the commodification of holiness, most famously Martin Luther's complaints about indulgences. In his Ninety-Five Theses, Luther argued that by allowing people to purchase salvation, the true meaning of the Gospel is corrupted. The "treasures of indulgences are nets with which one now fishes for the wealth of men," he writes, a statement that applies equally to the treasures of "nature" sold by Chopra and Paltrow. Luther saw the sale of indulgences for what they had become: a money-making scheme that exploited genuine anxiety about the afterlife to generate money for nobles and the pope. In a secular world, the fear is not about being unsaved but being unnatural—and we should be wary, as always, of those who promise the ability to purify yourself through purchasing power.

In a less well-known example of the same principle, during the nineteenth century a controversy broke out in New England over a practice known as "pew rental." For centuries, pews in many Christian churches could be privately owned. Choice seats belonged to wealthier families that held them across generations, their bloodlines turning the seating arrangement into a passable map of social hierarchy. The rich were literally closer to God's word, and the men delivering it were more likely to see them than the poor folks in the back. Eventually, cash-strapped American churches starting renting pews to secure regular income, and some spiritual leaders observed that the practice led to adverse effects on churches' priorities. "Rich men have become necessary to us," complained B. T. Roberts, a prominent Methodist bishop in New York. He saw fellow Methodists consumed with desire for social prestige, and church leaders more concerned with wealth than the word. "There is not always joy in the Church on earth over every sinner that repenteth, but over the rich sinner coming into the Church there is great rejoicing," he noted ruefully.[15]

Another way to think about Roberts's complaint is in terms of the relationship between holiness and the community. Roberts wanted Christian institutions to focus on redeeming the entire human community. Pew rentals worked against that focus, symbolically framing holiness as a privilege accessible only to specific individuals. "Houses of worship should be, not like the first class car on a European railway, for the exclusive, but like the streets we walk, free for all," he preached. "Their portals should be opened as wide for the common laborer, or the indigent widow, as for the assuming, or the wealthy."[16]

In her article "Who Deserves to Be Healthy? The Prosperity Gospel according to Goop," for *America*, a Catholic magazine, Eloise Blondiau suggests that the prosperity gospel in any form cultivates an unhealthy sense of entitlement and an exaggerated vision of our own control over fate.[17] It requires us to "suspend disbelief" about the real causes of income inequality and health disparities. Morality and metaphysics end up serving a business model, which runs the risk of distorting both. When they are conscripted into condo and candle sales, natural and nature are bound to look a little different. The wealthy target audience doesn't want to hear that purity can be had on the cheap. Neither do the salespeople, whose livelihood depends on the opposite. And if the rituals that secure natural purity happen to reinforce a socioeconomic spiritual hierarchy, that's not a concern—even if it adds another indignity to the state of those less fortunate.

Abandoning the prosperity theology of natural goodness means thinking critically about the assumptions that underpin it: "natural" as a synonym for "holy"; natural purity as a state secured by individuals, not communities; and the effectiveness of purity rituals as a function of their price.

If natural is not holy, then consecrated consumption does not expiate us from the sin of participating in a system that threatens the environment. Securing Mother Nature's approval by purchasing things with her name on the label is no more reasonable than saving one's soul with a payment to a priest. To the contrary: high socioeconomic status makes one's ecological footprint far greater—larger homes to air-condition, longer flights for vacations—and the effects of being "green" or "clean" are minimal. Reporting in *Vox* on a study conducted in Germany, David Roberts puts it well: "Rich people emit more carbon, even when they recycle and buy canvas tote bags full of organic veggies."[18]

This truth is obscured when rich people are the ones with time and resources to spend on living naturally. "People think environmentalists are the ones driving EVs [electric vehicles], but forget about the twenty people riding the bus," remarked California's attorney general Xavier Becerra in 2018.[19] Part of the reason we forget is that consecrated consumption forges a strong, mistaken association between wealth and naturalness.

The same association also makes us forget the real reasons that wealth leads to health. Reading Goop, you'd be forgiven for thinking that rich people are kept alive by a life full of expensive natural products. Everything clean and pure and free of artificiality—no wonder wealthy Americans can live up to fourteen years longer, on average, than the poorest Americans. But in truth, ultra-natural mascara does as little for one's personal health as tote bags of organic veggies do for the environment. The real health benefits of being wealthy are much less romantic: better access to education and health-care; lower rates of smoking, heavy drinking, and drug abuse; less stress; better exercise patterns and eating habits.

By exaggerating the benefits of circadian-rhythm condo lighting and naturally sourced homeopathic candles, Chopra and Paltrow distract from expert consensus on strategies that will improve everyone's quality of life, strategies that are necessarily communal and affordable across income brackets. Joe Colistra, a professor of architecture at the University of Kansas, expressed deep skepticism about the luxury approach to wellness building design. "What we're looking at are population-health strategies in which health and wellness is accessible to everyone," he told the *Atlantic*'s James Hamblin. "When you're talking about $15 million condos, it very quickly devolves into social inequities. Health is divided between the haves and have-nots."[20] Colistra emphasized thinking about outcomes in terms of communities across generations, not individuals syncing their rhythms with nature: "What's primary to health is social connectivity. Technology is secondary to community. So the goal is to create neighborhoods where people can thrive at all stages of life—life-long neighborhoods with great parks and schools and transit. Affordable inter-generational living is hard to come by.

And that's where healthy living fundamentally departs from the condos you mentioned. Those are almost segregated living environments."

Colistra is not alone in his concerns. Sustainable design expert Simona Fischer leveled precisely the same critique in a 2017 review of the wellness standard employed in Delos condos. She pointed out that the "solutions"—even if they are effective, which is by no means clear—can't be affordably scaled and therefore won't be implemented in schools and affordable housing. If there really are problems with air quality and water purity, it's likely those problems affect all members of the neighborhood. Solving them with expensive filtration systems is a mistake. "Instead of installing end-point water filters on taps and in showers and requiring testing at the tap, what if the design included an option to contribute funds toward water filtration at the neighborhood or city source, or help finance improved regular city-wide water testing at the water treatment plant?" suggests Fischer. "A system-wide solution would contribute more to health and wellness than providing perfectly purified water for a select group of occupants (who, if they are not poor, are statistically less likely to be experiencing a high body burden of toxicity than their lower-income counterparts anyway)."[21]

The same line of critique applies to expensive natural products. Natural prosperity theology tells us to pursue purity by purchasing from the high priests and priestesses of Nature. For those who can afford it, there's empowerment in saving the world one Church-scented candle and bar of "good" soap at a time. But this approach is neither workable for most people, nor scalable so that it could be. Buying lots of crystals like the amethyst built into Goop's $84 "crystal-infused" water bottle—may make you feel like you are aligning yourself with nature, but that doesn't mean you are saving it. Asked about the sourcing of these crystals by Emily Atkin of the *New Republic*, Goop refused to comment. This is understandable, writes Atkin, given that crystals are often "mined in countries with notoriously lax labor and environmental regulations, and some came from large-scale U.S. mines that have contaminated ecosystems and drinking water."[22]

Buying a bottle of Tourmaline Spring water gets you closer to water from a natural source in an unprocessed form. But it also needs to be bottled and transported, a trade-off that makes it a bad choice if you care about your ecological footprint. It's true that—as the advertising copy reminds us—buying a Glacce crystal bottle from Goop to carry around your drinking water will cut down on plastic waste. But drinking tap water from a regular old glass bottle does the same thing—and anyone can afford it.

None of this should lead to the reactionary mistake of labeling wealthy people as inherently impure, or concluding that ecologically conscious products are a complete waste of money. Rather, it should inculcate a healthy skepticism of natural purity when the definition of purity reflects and reinforces socioeconomic privilege. In addition, we should be cautious about solving problems with our relationship to the natural world by focusing only on what we buy. There's nothing wrong with buying expensive natural products and living in a purified condo. It's just that these are aesthetic preferences, not purifying rituals. They do not belong to a hierarchy of natural goodness, no matter how much the word "natural" gets folded into them. Insisting otherwise is no better than aligning with the Brahmins who saw their religious rituals as a source of exclusive divine purity—a view that, unlike the rituals themselves, is genuinely harmful. In the words of Martin Luther's thirty-second thesis: "Those who believe that they can be certain of their salvation because they have indulgence letters will be eternally damned, together with their teachers."

PART III

Law

M y daughter has three hermit crabs: Tittle-Tooth, the smallest; Giraffe,
the bully; and Rosebud, the shy one. After placing them in their ter-
rarium, luxuriously appointed with cholla wood for climbing and moss for
burrowing, we watched, fascinated, as Giraffe fought Rosebud for her shell
and Rosebud chirped in protest. We rooted for Tittle-Tooth as she (or he?
it's very difficult to sex a hermit crab) managed to claw her way into the big
water dish and then claw her way out again. Though they technically belong
to my daughter, our whole family has been enchanted, especially my wife,
who lovingly supplements their hermit meal with grapes and popcorn. My
favorite nature video is now a five-minute segment about hermit crabs nar-
rated by David Attenborough, in which at least a dozen line themselves up
according to size before performing an intricate, you'll-only-believe-it-if-
you-see-it shell-swapping ballet.

So it was distressing to find these wonderful creatures slandered in an influential book, *Natural Law in the Spiritual World*, by the biologist and theologian Henry Drummond. Published in 1883 and reprinted for decades, the
book contains a section that accuses hermit crabs of failing to live by nature's
laws, which, for Drummond, are also God's laws. Hermit crabs were "meant
for higher things," but instead of behaving as "perfect crustaceans" should,
they got lazy, borrowing mollusk shells to secure their safety. Their action
is a "twofold crime" that results from moral degradation: "First a disregard

of evolution, and second, which is practically the same thing, an evasion of the great law of work."[1]

It's because of this moral degradation, Drummond argues, that hermit crabs look so weird outside of their shells. They've been punished by Nature with a diminished physiology. "To the eye of Science," he writes, the hermit crab's "sin is written in the plainest characters on its very organization." Humans can learn from the degraded state of these pitiful decapods because nature's laws apply to every species: "The spiritual principle to be illustrated in the meantime stands thus: *Any principle which secures the safety of the individual without personal effort or the vital exercise of faculty is disastrous to moral character.*"

The political implications of this spiritual principle are clear enough. If, through government policy, citizens are supplied with safety and shelter, they run the risk of being degraded morally and physically, just like hermit crabs. The same principle can be used to distinguish good religion from bad. "Roman Catholicism offers to the masses a molluscan shell," says Drummond dismissively. "They have simply to shelter themselves within its pale, and they are 'safe.' But what is this 'safe'? It is an external safety—the safety of an institution." The laws that govern society and religion should reflect the laws that govern nature, a divine book of moral principles that reward virtue and punish vice.

Today Drummond's approach seems a bit outlandish because it completely collapses two different sets of laws, *descriptive* and *prescriptive*, that we are used to keeping separate. Descriptive laws are the domain of science, and deal with regularities in the world of facts. They describe what *is*. Prescriptive laws are the domain of politics, ethics, and religion, and describe what we *ought* to do. The evolutionary biologist Stephen Jay Gould used the term "nonoverlapping magisteria" to describe the division between scientific authority and moral authority, and thus the scope of their respective laws. Science, for him, is about discerning factual truths, whereas religion and secular ethics deal with normative truths. In 1999, the National Academy of Sciences affirmed Gould's position: "Science and religion occupy two separate realms of human experience. Demanding that they be combined detracts from the glory of each."[2] On this understanding, any resemblance between the two realms is merely an artifact of language.

Gould's position is a radical departure from Drummond. And not just Drummond: some version of the unity of natural law and moral law has been central to philosophical and religious thought for millennia. It is found in Eastern and Western thought, in Aristotle and at the core of modern democratic principles. Consider the first paragraphs of the Declaration of Independence:

> When in the Course of human events it becomes necessary for one people to dissolve the political bands which have connected them with another and to assume among the powers of the earth, the separate and equal station to which the Laws of Nature and of Nature's God entitle them, a decent respect to the opinions of mankind requires that they should declare the causes which impel them to the separation.
>
> We hold these truths to be self-evident, that all men are created equal, that they are endowed by their Creator with certain unalienable Rights, that among these are Life, Liberty and the pursuit of Happiness.

In his understanding of the relationship between natural rights and the laws of Nature's God, Thomas Jefferson followed the leading political philosophers of the day. He drew in part on the thought of Jean Jacques Burlamaqui, a Swiss "professor of Natural and Civil Law"—an official title that unites the two realms. Burlamaqui argued that "the law of nations is of equal authority with the law of nature itself, of which it constitutes a part, and [...] they are equally sacred and venerable, since both have the Deity for their author."[3]

Modern champions of so-called natural law theories lean on arcane terminological distinctions to argue that philosophers like Burlamaqui never confuse the *is* of the natural world with the *ought* of morality. They think that for Jefferson and Burlamaqui the word "natural," as it pertains to humans, means something like "rational," and point out that "nature" has not always meant "natural world" as it does today. Natural law theories supposedly employ the term in a different sense, more like the way it was used by ancient Greek philosophers and Christian theologians. Dismissing natural law by accusing it of committing the "appeal to nature fallacy" fails to take this history into account. There is no fallacy, it is claimed, once the "natural" in natural law is defined as "discernable through reason," and "nature" as describing the "essence" or "telos" of something.[4]

But as Drummond's massive bestseller makes clear, and as the following chapters will show, this was not at all how "natural" and "nature" functioned in the popular consciousness. Moreover, philosophers and theologians often failed to adequately distinguish one sense from the other, routinely conflating the two. Natural law did not simply mean "law discernable through reason" or "law pertaining to the essence of things." The very idea of "natural laws" ordained by God invites haphazard conversions of the *is* of the natural world's regularities into the *ought* of politics and religion. If God ordered what is, then the order of what is must be *good*. The laws that govern the activity of planets and chemicals and creatures like the hermit crab are infused with morality.

Take Constantin François de Chassebœuf, comte de Volney, a friend of Jefferson's whose work Jefferson admired enough that he translated portions of it. Volney was explicit about how, to use Gould's term, the "nonoverlapping magisteria" actually overlapped. "It is a law of nature that the sun illuminates successively the surface of the terrestrial globe," wrote Volney. That law, like all scientific laws, is part of "the constant and regular order of facts, by which God governs the universe; an order which his wisdom presents to the senses and reason of men . . . to guide them, without distinction of country or sect, toward perfection and happiness."[5]

Volney even includes a numbered list of the attributes possessed by the law of nature, which mixes descriptive terms with moral ones, just as you would expect if nature were synonymous with God:

> 1. Primitive; 2. Immediate; 3. Universal; 4. Invariable; 5. Evident;
> 6. Reasonable; 7. Just; 8. Pacific; 9. Beneficent; and 10. Alone sufficient.[6]

As a consequence of this conflation, it follows that every feature of the natural world should have an explanation that aligns it with the perfection and happiness of man. To take but one example: an extremely popular eighteenth-century book called *Nature Delineated*, written by a French priest, explained that corn is more difficult to grow than flowers because "the Divine Wisdom" wanted humans to associate daily subsistence with laborious care, and agreeable amusements such as flowers with easy recreation. The plants' design signals their relative utility to humans, and governs our proper attitude toward them.[7]

When Henry Drummond critiqued hermit crabs, he did so based on a widespread understanding of natural law that saw moral lessons inscribed in the natural world, designed by a benevolent God to punish sinners and reward the virtuous, whether humans or crustaceans. Even if natural law theorists before and after him claim to define "natural" as "rational" or "essential," their belief in a benevolent, omnipotent deity frequently leads them to see the laws of the natural world—in the colloquial sense of "the world that was not designed by humans"—as forces for good, organized according to benevolent divine principles.

The idea that humans should follow the laws of these natural systems is still very much with us. In a 2018 interview the US senator Ben Sasse blamed the artificial technology of modern social media for creating dangerous ideological polarization, as opposed to previously healthy divisions that emerged organically. "I think political tribalism is ramping in our time, because there is this collapse of local tribes, good tribes, natural tribes, traditional tribes," he said.[8] There is also a persistent sense that nature—again in the colloquial sense—punishes sin. In the face of natural disasters, for instance, some religious leaders blame immorality. In an interview that bemoaned increasing acceptance of abortion and homosexuality, televangelist Jim Bakker called Hurricane Harvey God's "judgment on America." His interviewer agreed: "The real issue with the weather and everything else on the earth has to do with sin and wickedness."[9]

While natural rights are typically associated with the equality of all humans, appeals to nature's laws have also been used to argue for the natural superiority of a few. "We were not meant to beg for moral validation from some of the most despicable creatures to ever populate the planet," declared the white nationalist Richard Spencer in 2016. "We were meant to overcome—overcome all of it. Because that is natural and normal for us. Because for us, as Europeans, it is only normal again when we are great again."[10]

Separating the laws of nature from moral laws, as Gould does, may seem like an effective way to shut down Spencer's naturalistic racism and the lunacy of blaming natural disasters on "sexual perversion." But in truth, it's only because the spheres *do* overlap that we can dismiss those claims. The *is* described by science affects the plausibility of *ought* statements. If natural disasters really did strike cities that support gay marriage with lawlike regularity, it might make sense to be cautious about gay marriage. The association has no empirical basis, however, so the moral principles that appeal to it lose their justificatory force. *Is* defuses *ought*.

Gould's neat separation is an impossibility. The regularities of the natural world do influence how we govern ourselves. No, hermit crabs are not a divine moral lesson in self-sufficiency. Corn was not designed to teach us the value of hard work. But that doesn't mean the *ought* of moral laws can be fully separated from the *is* of nature's laws. The question is not whether the two are related, but rather how we should understand their proper relationship once nature is no longer deified.

CHAPTER 7

The Invisible Hand

The growth of large business is merely a survival of the fittest. . . . The American Beauty rose can be produced in the splendor and fragrance which bring cheer to its beholder only by sacrificing the early buds which grow up around it. This is not an evil tendency in business. It is merely the working out of a law of nature and a law of God.

—JOHN D. ROCKEFELLER JR.

It is clear that the class struggle of the proletariat is a quite natural and inevitable phenomenon.

—JOSEPH STALIN

TO COMMEMORATE HIS FIFTIETH BIRTHDAY on Instagram, Dr. Shawn Baker shared a shirtless selfie alongside a photo of his cake. Well, not a cake: Baker is a strict carnivore, so to celebrate he seared a slab of ribeye (his typical diet is four pounds of steak per day), draped it on a pile of bacon, and staked the whole thing with a "5" and a "o" candle.

A talented athlete who excels at short distances on the rowing machine, Baker is playing a key role in popularizing carnivory, which has now gained enough of a cultural foothold to appear in major newspapers and secure a few celebrity endorsements. The diet is extreme. Not just low-carb; zero-carb. Not mostly meat; only meat. Though some go carnivore in an attempt to deal with chronic conditions, evangelists like Baker maintain it is the ideal diet for all humans. "We Evolved As Apex Predators," declares his website, a statement that is simultaneously scientific truth and life philosophy. As with advocates of virtually any diet, many carnivores argue that their approach is natural, a dietary route to paradise past. "We only switch to omnivore when we're starving or if we have a famine," says Travis Statham, a well-known figure in the carnivory world. On the carnivore diet, he believes, the human body can achieve its "primal perfection state."[1]

Strong beliefs about what constitutes a natural diet usually overlap with strong beliefs about physical health and the natural world. Paleo dieters end up in the CrossFit gym; vegans tend to support animal welfare. So at first it may seem utterly bizarre to discover that carnivory enjoys outsize popularity in a subculture that has nothing to do with health or nature: cryptocurrency enthusiasts. "Before there was money, before there was gold, there was meat! The original bitcoin!" tweeted Baker in 2017, to promote an episode of his podcast that explored the intersection of meat-only diets with a passion for decentralized digital currency. The paradox in his use of the word "before" is striking. Like all diets that lay claim to naturalness, carnivory adopts a justificatory rhetoric of returning to a better time in the past. How could a virtual currency built on high-tech cryptography and "mined" by computers possibly make the same kind of claim? What could be natural or primal about bitcoin?

The answer lies in recalling the sense of "natural" as "free from human interference." On Baker's podcast, the economist, carnivore, and author of *The Bitcoin Standard*, Saifedean Ammous, makes the case that eating vegetables, like using state-sponsored currency, is the result of artificial government manipulation. "In the same way that the US government and most governments in the world have been telling people what to eat . . . in economics there's a very similar problem," explains Ammous. "Instead of money being a free-market institution, which is what we had under the gold standard, [now we have a] twentieth-century model of government telling you no, you have to use this piece of paper."[2] Humans naturally gravitated to gold in multiple cultures, points out Ammous, just as they naturally gravitate toward eating meat. He believes that our current choices, of food and currency, reflect artificial constraints imposed by institutions that don't have our interests in mind.

Cryptocurrency returns us to an idyllic time when governments didn't control our currency; carnivory does the same with our diets. It's a match made in paradise past. "We live in a digital world, so we can't transact with gold, but Bitcoin has all the properties of gold, which was the earliest form of currency that we know of," said Ferdous Bhai, another bitcoin carnivore, in an interview for *Vice* magazine. "The philosophies and principles behind

Bitcoin are ancient."[3] In ancient times our dietary habits and economic habits emerged naturally, and by virtue of that natural origin they were superior to what we have today. As the statistician and best-selling author Nassim Taleb—not a carnivore but a Paleo dieter—enthuses in his introduction to Ammous's book: Bitcoin is "the first organic currency."[4]

The underlying model used to connect carnivory and bitcoin is an evolutionary one, where the wisdom of allowing evolution to happen naturally results in perfection. It's a model that animates much of the libertarian ethos found in Silicon Valley, an ethos that has long influenced government policy. As Arthur Levitt, the former chair of the US Securities and Exchange Commission, put it in a 2001 speech at Stanford: "The SEC's objective or function is not to dictate a particular market model, but rather, to allow the natural interplay of market forces to shape markets according to the demands of investors."[5] Organic evolution is good. Artificial systems are bad.

This understanding of naturalness collapses natural laws into political laws. The *is* of nature becomes an *ought*. When the individuals operating the cryptocurrency exchange platform Luno praise cryptocurrency as the "natural evolution of money," they are trading on the idea that evolution leads to better things: "It is not perfect, and will always continue to evolve."[6] When Ammous argues, in *The Bitcoin Standard*, that currency should be determined by the "natural working of the market system," he repeats a common justification of his preferred economic system that dates back to the time of Plato: whatever system is most natural is best.[7] In the language of professional theologians and philosophers, the natural evolution of market systems is "teleological," serving the purpose of improvement.

While the history of economic theory is filled with justificatory analogies to natural systems, the theories do not always refer to the same natural system. They can't, because the natural world is not a unified system that can be described according to a single model. Nature contains many systems, described by different branches of the natural sciences. The laws of planetary motion are not the laws of evolution. Each has its own vocabulary, its own agents and forces, its own data. Unarticulated in evolutionary justifications of bitcoin are the reasons for choosing evolution as the natural model that most accurately describes economic phenomena, much less determines how they ought to be. Why evolution instead of the body's circulatory system, or planetary motion, or hydrodynamics—all of which have been used to model the economy?

The next section of this chapter provides an answer, albeit one that deals a blow to justifications of economic policies based on appeals to nature. Good scientific models should map rigorously onto the target system, with clear criteria for how they do so. This is generally not the case with natural models of economic systems. Throughout history, the choice of the natural model used to advocate for economic policies has had its basis in cultural significance, intuitive plausibility, and persuasive power. Bitcoin enthusiasts don't choose evolution as a model because evolutionary models are uniquely suited to describing the development of currency. Rather, they choose it because evolution is the scientific model that currently commands the most authority. It's entirely possible, of course, that natural systems can be used to accurately and usefully model economic systems.[8] This chapter seeks to show only that in the majority of cases they are not—and, consequently, we should be suspicious of the ideological conclusions that follow from the natural model du jour.

In ancient Greece, political and economic systems were organized to take advantage of people's intrinsic natures. "Our several natures are not all alike but different," remarks Socrates in Plato's *Republic*. "One man is naturally fitted for one task, and another for another."[9] A basic version of this approach is uncontroversial. Very young children clearly have different capacities than adults, and we assign their respective civic duties—and legal rights—accordingly. Division of labor also respects differing natural capacities. Some people have better vision, others better hearing, and it would be strange to task a blind person with directing traffic, or a deaf person with evaluating a music competition.

But discussions of natural capacities and the division of labor they ought to entail quickly stray into controversial claims. "Is there anyone thus intended by nature to be a slave and for whom such a condition is expedient and right," asks Aristotle in the *Politics*, "or rather is not all slavery a violation of nature?" He answers his own question with a defense of slavery. "For that some should rule and others be ruled is a thing not only necessary, but expedient; from the hour of their birth, some are marked out for subjection, others for rule." Put simply, to honor the different "natures" created by nature is to approve of slavery. It is clear that "some men are by nature free, and others slaves, and that for the latter slavery is both expedient and right." This conclusion was simultaneously biological, metaphysical, and economic—certain humans are suited by nature to be owned—and it also served as a legal justification for slavery.[10]

To strengthen his case, Aristotle leans heavily on a natural analogy: the state as human being. He borrows it from Plato, who believes that the "best governed state" is most like "an individual man."[11] Plato models the proper division of labor among people on the proper division of labor among different parts of the body. Expanding on the analogy, Aristotle states that all people have a body and a soul. Since the soul is superior to the body, "of the two, one is by nature the ruler, and the other the subject." This hierarchical duality is at once natural (nature gives us a body and a soul) and political (one is the ruler and the other a subject). The model of the body is a model for society.

Aristotle sees the natural duality of ruler and subject, of soul and body, everywhere: in the proper relationship between humans and animals, and between men and women: "The male is by nature superior, and the female inferior; and the one rules, and the other is ruled." The reasonableness of slavery proceeds from the application of the same principle, and he observes that among humans, "the lower sort are by nature slaves, and it is better for them as for all inferiors that they should be under the rule of a master."[12]

Long after Aristotle, the model of organizations as human bodies shapes everyday language. Corporations take their etymological root from *corpus*, Latin for "body." For a territory to become part of a country it must be incorporated, its citizens grafted onto the "body politic." The soul, secularized, becomes the brain. Presidents, prime ministers, and monarchs are referred to collectively as heads of state, brains ruling over the body of the citizenry.

This is no mere linguistic flourish. Plato, Aristotle, and other influential thinkers who followed them believed that it represented an important truth. A deified nature has authored reality such that the ideal human political system functions like a body, and you can deduce the former by looking at the latter. Departing from the model of a body is unnatural, and it will do the body politic harm just as it will do an organism harm. In their arguments, the model of "X as body" works—or is meant to work—the same way that a mathematical model of a bridge would in arguments about how to build bridges. You can build many different types of bridges, but if they depart from the mathematical model of a stable bridge they will collapse. Likewise, departing from the model of a stable and properly ordered human body will cause a state to collapse.

Since the economy is part of a state, and states are modeled on human individuals, it follows that nature's intentions for economic policy can also be read off the model of a body. In *The Wealth of Nations*, the enormously influential philosopher and economist Adam Smith does exactly that, mapping commerce onto the circulatory system to make a case against excessive government regulation of trade routes. Goods and capital are the lifeblood of a nation. Allowed to flow naturally, the body's circulatory system functions well, distributing that lifeblood as it should, keeping the organism healthy. But, as Smith argues in a vivid passage, a badly managed state can thwart the natural circulatory system. Great Britain's laws have "broken altogether that natural balance which would otherwise have taken place" in an economy. The result is disastrous:

> Her commerce, instead of running in a great number of small channels, has been taught to run principally in one great channel. But the whole system of her industry and commerce has thereby been rendered less secure; the whole state of her body politic less healthful than it otherwise would have been. In her present condition, Great Britain resembles one of those unwholesome bodies in which some of the vital parts are overgrown, and which, upon that account, are liable to many dangerous disorders, scarce incident to those in which all the parts are more properly proportioned. A small stop in that great blood-vessel, which has been artificially swelled beyond its natural dimensions, and through which an unnatural proportion of the industry and commerce of the country has been forced to circulate, is very likely to bring on the most dangerous disorders upon the whole body politic. [...] The blood, of which the circulation is stopt in some of the smaller vessels, easily disgorges itself into the greater, without occasioning any dangerous disorder; but, when it is stopt in any of the greater vessels, convulsions, apoplexy, or death, are the immediate and unavoidable consequences.[13]

Overzealous management of the free market interferes with Smith's famous invisible hand, which naturally orders commerce in the same way that it naturally orders the human body. For Smith, undesirable economic regulations represent the same dangerous meddling with natural systems that surgery and "artificial" medicine do for someone who believes in natural healing. Indeed, he uses precisely that analogy:

> The uniform, constant, and uninterrupted effort of every man to better his condition, the principle from which public and national, as well as private opulence is originally derived, is frequently powerful enough to maintain the natural progress of things towards improvement, in spite both of the extravagance of government, and of the greatest errors of administration. Like the unknown principle of animal life, it frequently restores health and vigour to the constitution, in spite not only of the disease, but of the absurd prescriptions of the doctor.[14]

For Smith there's no question that the invisible hand organizing the economy works in our best interest. Nature's laws not only order reality, but that order is beneficial, in his words a "natural progress of things towards improvement." This is another way of saying that Smith's understanding of naturalness is inherently theological, which, at the time, would have been entirely consistent with the perspective of most influential thinkers, even those who, like Smith, were not especially religious. Throughout the mid-twentieth century, secular admirers of Smith downplayed the theological aspects of his philosophy, dismissing them as irrelevant or merely rhetorical. They were aided in this task by Smith's general reliance on natural theology and the vocabulary of "nature" instead of explicit appeals to the Christian God of his contemporaries. But God disguised as nature is still God, and the political theorist Lisa Hill has collected overwhelming evidence for its centrality to Smith's thought. He refers to the "beneficent ends which the great Director of Nature intended to produce," the "benevolent purpose of Nature," and contends that "the happiness of mankind" is in fact "the original purpose intended by the author of Nature." There can be no question, concludes Hill, about Smith's belief in "the miraculous order of Nature," and the importance of that belief to his economic theories.[15] Faith in the invisible hand is, indeed, a faith.

Similar rhetoric can be found in the work of Smith's friend, the political theorist Edmund Burke. Though Burke recognized the need to depart from a "state of nature," at times he also sought to ground his political and economic principles in the theological wisdom of natural processes. Writing against government subsidies, for instance, Burke warned about the hubris of those who would "supply to the poor those necessaries which it has pleased the Divine Providence for a while to withhold from them. We, the people, ought to be made sensible that it is not in breaking the laws of commerce, which are the laws of Nature, and consequently the laws of God, that we are to place our hope of softening the Divine displeasure to remove any calamity under which we suffer."[16] All that's missing is a discussion of hermit crabs.

However appealing it may be in theory, the benevolent design of Nature rarely works out in practice, requiring intellectual acrobatics on the part of those who invoke it. Smith recognizes that a healthy economic circulatory system depends on some government interference. Complete freedom leads to monopolies, giving manufacturers outsize power over prices and politicians, which works to the detriment of the body politic. How to account for monopolies while maintaining an ideal of naturalness? Just call them unnatural. Monopolists, writes Smith, are guilty of selling their commodities "much above the natural price." To regulate them is to force them into accordance with nature—even though monopolies themselves naturally emerge in unregulated economies.

It's not just accounting for the imperfections of natural economies that reveals internal inconsistencies. Attempts to map natural models onto cultural systems tend to break down under any real scrutiny. Take the analogy of political states to individual organisms. Here is Jean-Jacques Rousseau doing his best to make it all hang together, and failing:

> The body politic, taken individually, may be considered as an organised, living body, resembling that of man. The sovereign power represents the head; the laws and customs are the brain, the source of the nerves and seat of the understanding, will and senses, of which the Judges and Magistrates are the organs: commerce, industry, and agriculture are the mouth and stomach which prepare the common subsistence; the public income is the blood, which a prudent *economy*, in performing the functions of the heart, causes to distribute through the whole body nutriment and life: the citizens are the body and the members, which make the machine live, move and work; and no part of this machine can be damaged without the painful impression being at once conveyed to the brain, if the animal is in a state of health.[17]

There is nothing scientific or rigorous about Rousseau's attempt to superimpose natural models onto political economy. The supposed associations are not evaluated according to anything like a scientific standard of accuracy. When Rousseau says the citizens are "the body and the members," he doesn't specify whether they are the limbs of the body (Plato uses the example of a finger) or cells in the body. In practice this is a crucial detail, insofar as it determines the relative disposability of individual citizens to the body politic. A few skin cells lost? No problem. A finger or an arm? Best to protect those citizens. The points of homology between a natural model and a political system need to be specified, because legislative consequences depend on them. Is the "head" of state more like a democratically elected president or a sovereign? Do the laborers who work to prepare the king's food count as part of the head, or the arms? Does the circulatory system include physical goods and money? Are citizens objects of value, thus making up part of the blood, or are they the flesh nourished by the blood? There are no good answers because the model was chosen with an eye to rhetorical power, not accurate correspondence.

Switching to other natural systems does not help. Every new development in the natural sciences has spawned new attempts to explain cultural phenomena with universal models, and all have suffered serious shortcomings. When Isaac Newton revolutionized science in the late seventeenth century, many intellectuals believed his mathematical models of the physical world disclosed universal principles that governed the entire cosmos—principles that were uniformly understood as ordained by some kind of higher power, and therefore conducive to harmony. The historian of science I. Bernard Cohen highlights how this led to platitudinous nonsense in the work of the philosopher and theologian George Berkeley, who tried to explain human associations with Newtonian dynamics. "A like principle of attraction" works "in the Spirits and Minds of men," wrote Berkeley. Humans form "communities, clubs, families, friendships, and all the various species of society" because they are drawn to each other by a kind of gravity. Moreover, "attraction is strongest . . . between those which are most nearly related." No doubt Berkeley would have disagreed with the adage that "absence makes the heart grow fonder" because it violates the laws of physics.[18]

Sometimes, confusingly, the justificatory models of nature are drawn from multiple areas of the natural sciences. Adam Smith mixed physiology with physics, alluding to Newtonian forces of attraction in natural markets along with his circulatory system. Ralph Waldo Emerson did the same in his own defense of free trade. The "basis of political economy is non-interference," Emerson declared. "Do not legislate. Meddle, and you snap the sinews with your sumptuary laws." The implied physiological metaphor ("sinews") is immediately followed by an extended appeal to physics: "The laws of nature play through trade, as a toy-battery exhibits the effects of electricity. The level of the sea is not more surely kept, than is the equilibrium of value in society by the demand and supply: and artifice or legislation punishes itself, by reactions, gluts, and bankruptcies. The sublime laws play indifferently through atoms and galaxies."[19]

The laws of physics remained a popular justificatory analogy for economic laws well into the twentieth century. (Harvard economist Lawrence Summers, in 2017: "The laws of economics are as difficult to defy as the laws of physics.")[20] But describing human interactions as if we are celestial bodies or particles is no longer in vogue. Newtonian metaphors have been eclipsed by Darwinian ones; not the pull of gravity, but the struggle to survive. Nowadays, the best way to secure Nature's stamp of approval for one's preferred economic system is by appealing to a comparatively new branch of the natural sciences: evolutionary biology.

Long before bitcoin carnivores talking about the evolution of natural currency, there was social Darwinism, a term as familiar as it is misleading. Typically associated with ruthless laissez-faire economics and racist breeding programs, for most people it conjures up the specter of unregulated robber barons crushing their competition and Nazi eugenicists trying to purify the German *volk*. Evolution can only take place in a brutally competitive natural world, which means that it's perfectly natural to exploit and eliminate weaker members of the species. Just as kings once claimed a divine mandate, the powerful can claim Nature's mandate, their superiority firmly established as scientific fact. Social Darwinists transform survival of the fittest into a philosophy of life, and into political and economic theories. Instead of sympathizing with the unfit, better to perfect society and its members by applying the natural principles discovered by Darwin.

There's some truth to this picture of social Darwinism. As Darwin's theories gained wider acceptance, "natural selection" was frequently paired with support for eugenics. The term "eugenics" was invented by Darwin's half-cousin, the British scientist Francis Galton, who drew inspiration from natural selection. Galton enthused about the prospects of breeding a better race, even drafting a utopian novel about a eugenics-based college. The novel went unpublished, but his ideas had lasting effects. One characteristic 1923 American high school biology textbook includes a section that proposes a eugenic solution to humanity's economic problems, drawing on Darwinian principles. "As soon as civilization began, we started to put a check on the operation of natural selection which in a state of nature allows the weakest to be destroyed," explains the textbook. Civilization encourages the "mentally enfeebled" to breed more than it does "mentally well-endowed people," resulting in high rates of poverty. The answer? Laws that "restrict the freedom of persons in whose blood there is proven to be the strain of inherited insanity or feeble-mindedness," and "prevent them absolutely from having offspring." With eugenics, civilization could reclaim the operation of natural selection and, with it, solve the problem of inadequate wealth distribution.[21]

It is also true that some influential laissez-faire capitalists embraced what they understood as Darwinian theory to justify their preferred practices. In his autobiography, the railroad and steel magnate Andrew Carnegie discussed how Darwin's *The Descent of Man* had helped him replace religious ideology with scientific truth. "Not only had I got rid of theology and the supernatural, but I had found the truth of evolution," explained Carnegie. For him, the truth of evolution held obvious implications for economics. Indiscriminate charity and government aid were ultimately aiding the weak and subverting nature's law of competition. He admitted that "the law may sometimes be hard for the individual," yet "it is best for the race, because it insures the survival of the fittest in every department."[22] Later, economists such as Milton Friedman used the metaphor of evolutionary theory to illustrate the virtues of free markets, in which economic agents left to themselves will naturally maximize returns. "The process of 'natural selection' thus helps validate the hypothesis," claimed Friedman, "or rather, given natural selection, acceptance of the hypothesis can be based on the judgement that it summarizes appropriately the conditions for survival."[23] His friend, the conservative thinker William F. Buckley Jr., believed in the same principles. The best economic policies, said Buckley, should accommodate "the natural desire of the individual for more goods" and allow the market to "take organic form."[24]

But as many historians have pointed out, this picture of social Darwinism (cemented by the historian Richard Hofstadter's polemical book *Social*

Darwinism in American Thought) is both too narrow and too broad.[25] Too narrow, in that it wasn't only laissez-faire capitalists and evil eugenicists who sought to justify political and economic policies with evolutionary theory. People across the ideological spectrum saw in evolution tacit support for whatever they happened to believe. And too broad, in that many of the people retrospectively labeled social Darwinists were not actually using Darwin's evolutionary theory. They simply did what people are still doing today, namely picking and choosing whatever biological principles happened to best fit their economic ideology.

Properly understood, what's now referred to as social Darwinism actually predates Darwin. Before *The Origin of Species* there existed multiple theories of reproductive fitness as it related to the "progress" of animal species, including humans. These theories were used to justify political and economic orders, as Thomas Paine did in the *Rights of Man*.[26] Paine needed to counter Edmund Burke, who, as we have seen, occasionally sought to ground his own political and economic principles in the laws of nature. "Aristocracy has a tendency to deteriorate the human species," Paine wrote. "By the universal economy of nature it is known, and by the instance of the Jews it is proved, that the human species has a tendency to degenerate, in any small number of persons, when separated from the stock of society. . . . The artificial NOBLE shrinks into a dwarf before the NOBLE of Nature."[27]

Here Paine builds his support for political revolution on the idea that aristocrats are naturally inferior. (There's also the passing mention of Jews made inferior by "unnatural" inbreeding, a racist trope popular among anti-Semites.) Any society that concentrates power in an inbred aristocracy is unnatural, unjust, and doomed to failure. Political economy should mirror the universal economy of nature, which, for Paine, represented a kind of divine will (if not the Christian one that he rejected). A better society will emulate the same natural principles that make for a better species.

After Darwin transformed the intellectual landscape, numerous egalitarian revolutionaries took inspiration from his theories. Their arguments against more familiar laissez-faire versions of social Darwinism paralleled the back-and-forth between Paine and Burke on the relative "naturalness" of aristocracy. One of the very earliest examples is *Le Darwinisme Social*, a short book by the French anarchist and revolutionary socialist Émile Gautier, published in 1880. Gautier acknowledged that at first glance, Darwin's theories appear to justify ruthless elimination of the weak. He quoted the renowned

German scientist Ernst Haeckel, who maintained that "any intelligent and informed politician should, it seems to me, recommend Darwinism as the best antidote against the absurd egalitarian theories of the socialists. . . . Darwinism is anything but socialist. . . . If one wished to ascribe it a political leaning it could only be called aristocratic."[28]

But Gautier thought this perspective misrepresented Darwin. "Far from being the most energetic condemnation of the socialist revolutionary," he argued, "Darwin's theory, if properly understood, rigorously observed, and taken to its logical conclusion, may, to the contrary, lend it some precious and unexpected competition." Like Paine, Gautier believed that concentration of power in a privileged elite does not occur "because nature so desired," but rather because unnatural human systems have allowed it to happen. "It is not by natural selection," insisted Gautier, "but rather by artificial selection that the weak are victims."[29] A natural distribution of material resources requires removing the artificial system that currently oppresses the masses—it requires a new set of laws and a new economy.

In Russia, another anarchist philosopher named Peter Kropotkin developed a similar vision of Darwin's theory in his book *Mutual Aid: A Factor of Evolution*. Kropotkin was an accomplished naturalist, and he drew on personal experience in his descriptions of migratory birds finding their way with "accumulated collective experience," and delicate fallow deer seeking strength in numbers. As a factor of evolution, Kropotkin thought sociability must count at least as much as competitiveness: "I saw Mutual Aid and Mutual Support carried on to an extent which made me suspect in it a feature of the greatest importance for the maintenance of life, the preservation of each species, and its further evolution."[30] Kropotkin moved seamlessly from his descriptions of mutual aid in the animal kingdom (chapters 1 and 2) to mutual aid in the human world (for example, chapter 3, "Mutual Aid among Savages," which includes a skeptical subsection on "the supposed war of all against all"). The book's evolutionary science, though meant to stand on its own, also served to validate Kropotkin's political and economic theory, a critique of capitalism and centralized government that favored communal exchange and small, organic communities.

Meanwhile, in China, Darwin's thought was appropriated by rival political factions. Embarrassed by a century of military defeats, Chinese thinkers were seeking a way to restore national dignity. "Reformers turned to Charles Darwin as a foreign authority on change," writes the historian James Pusey, "presenting him not first and foremost as a natural scientist who had discovered an amazing fact of life, but as a political scientist who had discovered a cosmic imperative for change."[31] On one side of the debate over Darwin were the gradualist reformers—among them the first translator of Darwinian ideas into Chinese, a man named Yan Fu—who thought Darwinian evolution progressed incrementally, and therefore political revolutions went against nature's laws. On the other side were the revolutionaries, including Sun Yat-sen, who believed that Darwinism validated rapid regime change. Like Haeckel and Gautier, everyone saw their preferred ideology confirmed as true and good in the mirror of "Darwinism." And despite their differences, everyone agreed that Darwin's theory described not merely biological change, but improvement. Pusey documents how subtle translation mistakes contributed to this position: "'Natural selection' came out as 'natural elimination'; the 'survival of the fittest' became the 'superior survive and the inferior are defeated.'"[32] (The first Japanese translation of *Origin of Species* turned "evolution" into "the theory of progressive change.")[33]

Later in the twentieth century, communists sought to buttress their ideology with Darwinism. Mao compared his "Let a Hundred Flowers Bloom" movement to Darwin's theory of evolution, and used the language of evolutionary biology to argue that "socialism, in the ideological struggle, now enjoys all the conditions to triumph as the fittest."[34] Joseph Stalin invoked Darwinian themes when he spoke of nature as the scene of endless struggles between opposed forces, which produce an "onward and upward movement" of improvement from "lower to higher." The result was biological justification for revolution: "If the world is in a state of constant movement and development, if the dying away of the old and the upgrowth of the new is a law of development, then it is clear that there can be no 'immutable' social systems. . . . Hence the capitalist system can be replaced by the socialist system."[35]

But communists were not consistent when it came to the kind of evolutionary theory they preferred. Since their use of evolutionary models was ideological, not scientific, they adopted bits and pieces of whatever fit their ideology. Marx and Engels were happy with some aspects of Darwinian evolution, but also saw it as a threat to their vision of class solidarity and teleological history. Overall they, along with Stalin, favored Jean-Baptiste Lamarck. (Stalin criminalized Darwinian evolutionary theory and encouraged pseudoscientific Lamarckianism.) Lamarckian evolution was also beloved by some advocates of laissez-faire economics. Strangely, the world's most influential "social Darwinist," the British philosopher Herbert Spencer, firmly rejected Darwin. Though he coined the phrase "survival of the fittest," which was subsequently adopted by Darwin in the fifth edition of *Origin*, Spencer saw nature the way Lamarck did—as a "beneficent force" that rewarded struggle *in this life* by changing the essence of the organism. This essential change would then be passed on to the next generation, allowing Spencer to validate the idea that hard work is rewarded (and indolence punished) by appealing to the natural world. In an ironic twist, the great champion of social Darwinism was really a Calvinist Lamarckian, who, in the words of historian Howard Kaye, championed "a Protestant ethic of self-improvement dressed up in a new vocabulary of evolutionary science."[36]

Notwithstanding their enormous differences, economic appeals to evolutionary biology have certain common elements. From Marx to Andrew Carnegie, from eugenicists to bitcoin carnivores, they all assume that the natural world and human society are organized according to the same principles. Whether they explicitly invoke God or not, these principles are theological—even for supposedly secular communists, who fold religious teleology into their worldview by disguising it as natural history. Arguments over which economic system is most natural assume that the virtues of an economic system can be demonstrated by showing how it is modeled on nature, a benevolent force oriented toward improvement. It's a vision of nature so reassuring that even Darwin was tempted by it. "And as natural selection works solely by and for the good of each being," he declared in the conclusion to *The Origin of Species*, "all corporeal and mental endowments will tend to progress towards perfection."[37]

Darwin firmly rejected attempts to make evolutionary theory the handmaid of extreme laissez-faire economic theory. In a letter to the geologist Charles Lyell, he complained that his work was being hijacked by those who wished to prove "'might is right' and therefore that Napoleon is right, and every cheating Tradesman is also right."[38] He was right to bristle at the unscientific audacity of using his theory to argue for unregulated capitalism and elimination of the weak. Yet he should have been no less upset if the theory were used to justify a more appealing ethic of helping fellow humans. Kropotkin's focus on neglected aspects of mutual aid has been praised by modern evolutionary biologists (though there was some of it already present in Darwin), but it doesn't count as evidence in support of his economic theories. Natural selection may be able to explain the origin of species, including the human species, but that process of selection is entirely amoral. The relative naturalness of communism or capitalism tells us nothing about whether we are "meant" to adopt them.

In addition to being prescriptively empty, the metaphor of economics as evolution is, at the very least, descriptively problematic, in the same way that metaphors of the economy as a living organism or a system of celestial bodies are problematic. When "natural selection" happens in an economic system, does it select for the fittest corporations or the fittest individual human beings? What's the equivalent of a "gene" in an economy? If natural competition for resources occurs not only between individuals but also between species, then economic theory needs a parallel unit for "species." Should it be the "nation," or smaller units such as community or family? These questions can never be answered adequately, much less scientifically, because evolutionary models, like any models drawn from the natural world, do not provide the master key for decoding all of reality.

In 1879 the German biologist Oscar Schmidt decried anyone who apotheosized natural processes in the service of socioeconomic ideals: "They have set up, in the place of a personal God, a sort of infallible bugbear under the guise of an Omnipotent Idea. The whole thing is misty, mystic, supernatural, in no sense scientific, least of all is it a Darwinian explication of facts."[39]

Yet here we are in the twenty-first century, setting up the same infallible bugbear. A prominent public intellectual praises bitcoin as the first "organic" currency, as if that somehow makes it good. Articles are still being written about how capitalism is "unnatural," as if being unnatural makes something bad. An online investment dictionary explains that collusion is illegal "because it interferes with the natural market forces of supply and demand."[40] Through all of this, the premise that models from nature can be neatly mapped onto economic systems remains largely unquestioned, instead of being revealed for the theological claim that it really is.

While the religiosity of secular economic theory is often hidden in the language of nature, religious traditions have always used explicitly theological principles to justify claims about how economies should work. Nowadays, the practice of charging interest is standard, but for a long time it was considered a horrific sin. (Islamic law still forbids it, and sharia-compliant banks must earn money with different financial instruments.) Numerous religious treatises were written decrying usury, and running through them all was an accusation even more bizarre than the association of bitcoin and carnivory: Usury is sodomy.

It was once commonplace to call usury "the sodomy against nature." In the *Inferno*, Dante placed usurers and sodomites in the same circle of hell, lower than murderers. "Violence may be done against the Godhead / By denial in the heart and blasphemy / And by despising nature and her bounty," explains Virgil to the pilgrim in Canto XI. This is the sin of the sodomite and the usurer alike: "He scorns nature." Fifty years later, Pope Gregory XI confirmed Dante's extreme distaste for both practices. "In the whole world I believe there are no two sins more abominable than those that prevail among the Florentines," he said. "The first is their usury and infidelity. The second is so abominable I dare not mention it."[41] The "unspeakable sin," the abomination he dared not mention, was a well-known euphemism for sodomy.

The strained logic of the association between usury and sodomy turns on what constitutes a violation of natural reproduction. Sodomites, goes the argument, take that which is naturally generative—genitalia, sexuality, the reproductive act—and make it barren. Usurers invert this sin, taking that which is naturally barren—currency, which is no more than a symbol—and through their artifice cause it to reproduce. This reproductive metaphor began with Aristotle, who described charging interest as "the breeding of money," an "unnatural act" that "makes a gain out of money itself, and not from the natural object of it."[42]

Opponents of usury recognized difficulties with this logic. Farmers, for example, use technology to increase crop yields beyond their natural level. That would appear to make farming an unnatural act, causing the earth to breed beyond its intrinsic abilities. But farming was obviously moral, so tortured rationalizations were introduced to justify it. Farming is a form of "natural usury," argued one Christopher Jelinger in 1679, "whereby a Man deals with Mother Earth, which the holy Fathers in old time used by God's own Institution." Jelinger's argument parallels Adam Smith's approach to monopolies. Since monopolies are bad, Smith must label them unnatural; since farming is good, Jelinger must label it natural.[43]

Opposing usury by associating it with an "unnatural" sexual practice isn't just another example of appeals to nature gone wild in the realm of economic theory. It's a reminder that theologians, no less than their secular counterparts, are attracted to the authority of nature. They, too, need its objective secular authority to strengthen their religious claims. The *ought* of religious laws, like the ought of economic laws, needs to be grounded in the *is* of nature. "God says so" is a much stronger claim when followed by "and so does nature." And, as the case of "unnatural sodomy" reminds us, this is especially true when it comes to God's laws about reproduction and sexuality.

CHAPTER 8

The Rhythm

It is now quite lawful for a Catholic woman to avoid pregnancy by a resort to mathematics, though she is still forbidden to resort to physics or chemistry.

—H. L. MENCKEN, *Notebooks*, 1956

GOD CARES ABOUT SEX. Just after the biblical commandment not to murder there's the commandment not to commit adultery, which is illegal in many countries and several US states, including North Carolina, where courts frequently order the adulterer outside the marriage to pay large sums in punitive damages. Laws against homosexual activity, which remained on the books in the US until 2003, are still widespread globally and almost always have a religious genealogy.

Opposition to contraception is also associated with religion, particularly Catholicism. On a 2015 flight back to the Vatican after visiting the Philippines, Pope Francis fielded a journalist's question about his position on contraception, as he has many times in the past. An important issue for all Catholics, it's especially so in the Philippines, a Catholic-majority country where battles over birth control access and abortion (which is against the law) have raged for decades, alongside high rates of unintended pregnancy, maternal mortality, and a thriving black market for herbal abortifacients, often sold near churches.

In his answer, Francis reaffirmed the Catholic Church's opposition to birth control, grounded in the teaching that contraception is a mortal sin as articulated in Pope Paul VI's 1968 encyclical, *Humanae vitae*. But then he shocked reporters by suggesting that some people may nevertheless have a duty to limit their children. "God gives you methods to be responsible," said the Pope. "Some think that—excuse the expression—that in order to be good Catholics we have to be like rabbits. No."[1]

The Pope, understandably, appeals to God. But given the historical position of the Catholic Church and other opponents of artificial contraception, he could just as well have appealed to nature. The "methods to be responsible" that Francis referred to fall under the general category of what is popularly referred to as natural family planning (NFP), the only option for married Catholics who want to have sex but don't want to procreate.[2] Condoms, the Pill, IUDs, surgical sterility, voluntary withdrawal (*coitus interruptus* in the traditional lingo): all are impermissible and together classed as "artificial." But intercourse timed during periods of infertility is acceptable, because it is understood as compatible with nature's laws.

There's an air of timelessness to NFP, as if this God-given natural method has been around as long as nature itself. And in certain ways it almost has: versions of the so-called rhythm method and the idea of a "safe period" date back to at least St. Augustine. But God did not choose to reveal an *effective* method, founded on genuine knowledge of women's fertility, until much later. From its earliest implementation through the 1800s, physicians and clergy catastrophically misidentified the safe period, confusing it with the *most* fertile time in a woman's cycle. Advocates of timed abstinence praised it for being nature's own way, but those actually using it to avoid pregnancy were less enthusiastic. "Nature is like the letter of the law which faileth," complained one Mary Hallock Foote of Idaho in a letter to a friend, despondent after becoming pregnant on the rhythm method.[3] Unsurprisingly, other forms of birth control such as condoms steadily increased in popularity, concerning moralists who saw them as encouraging sexual licentiousness and a lack of respect for physical intimacy as it naturally ought to be.

Only in the 1920s did two scientists—Japan's Kyusaku Ogino and Austria's Hermann Knaus—independently discover the pattern of human ovulation as it relates to the menstrual cycle, allowing them to accurately identify infertile periods. Knowledge of their findings slowly spread, and in 1932 the Chicago physician Leo J. Latz took it upon himself to share it with the American public. He self-published *The Rhythm of Sterility and Fertility in Women*, which became an instant sensation, selling hundreds of thousands of copies. Alongside four cherubic baby faces, the dust jacket celebrated the virtues of this new approach: *Hygienic. Dependable. Practical. Ethical.*

For Catholics concerned about the heresy of endorsing contraception, it was quite literally a godsend. "Divine Providence has come to the assistance of mankind at critical periods by unfolding nature's secrets," wrote Father Joseph Reiner, a Jesuit priest, in his introduction to *The Rhythm*. "It seems to be doing that in the present crisis by enabling scientists to discover 'the rhythm of sterility and fertility in women.' The discoveries of Doctors Ogino and Knaus show us the way out of the difficulty, without a compromise of principle."[4]

The reference to Divine Providence unfolding "nature's secrets" is no accident. A key feature of Catholic support for the rhythm method has been that it accords with the laws of nature and, by proxy, with God's laws. Latz, a Catholic himself, devotes nearly a third of his book to a section called "Ethical Aspects." While God appears infrequently, "nature" and "natural" are on nearly every page. In the section's first paragraph, Latz chooses to cite a portion of Pope Pius XI's 1930 encyclical on marriage that does not mention God, but does emphasize the importance of harmonizing with nature: "Nor are those considered as acting against nature who in their married life use their right [to practice natural birth control] in the proper manner, although on account of natural reasons of time or of certain defects, new life cannot be brought forth."[5] (The word "natural" in "natural reasons of time or of certain defects" is clearly used in the colloquial sense, not the idiosyncratic sense of "essential" or "teleological" that natural law advocates claim informs papal reasoning. Likewise, the word "nature" in Father Joseph Reiner's introduction to *The Rhythm* is also used in the colloquial sense. The same is true in many instances of the use of "nature" and "natural" in religious discussions of sexuality, which invites ambiguity and confusion in all instances.) Voluntary abstinence on certain days, explains Pius XI, is allowable precisely because it is not "a violation of nature" and can be "made use of without doing violence to nature."

The importance of the natural/artificial dichotomy to Catholic teachings on birth control can hardly be overstated. In the section of *Humanae vitae* that addresses birth control, Pope Paul VI warns of "certain limits, beyond which it is wrong to go, to the power of man over his own body and its natural functions." One of those limits concerns our power over natural reproductive function, and at first it might seem like any effective birth control method, whether condoms or NFP, would exceed that limitation. Yet the encyclical explicitly approves "recourse to infertile periods," contrasting it with forbidden *artificial* methods: "In reality, these two cases are completely different. In the former the married couple rightly use a faculty provided them by nature. In the latter they obstruct the natural development of the generative process."[6]

The Church's condemnation of the Pill as unnatural dealt a devastating blow to Dr. John Rock, the Catholic scientist and physician who helped design and popularize it. Rock firmly believed the Pill was just as natural as the rhythm method—he called it "an adjunct to nature"—and this naturalness was a central plank in the case for its ethical legitimacy.[7] His reasoning was straightforward. The Pill used progestin, a naturally occurring hormone. Not only that, Rock and his collaborators had formulated the Pill with a week-long placebo period so that monthly menstruation could happen, not because this was medically necessary but rather, as Malcolm Gladwell puts it in an essay about the Pill's origins, "because Rock wanted to demonstrate that the Pill was no more than a natural variant on the rhythm method."[8]

In light of the popular association of the rhythm method and NFP with Catholicism—"Catholic roulette" as some derogatorily refer to it—one could easily mistake emphasis on naturalness for an inherently Catholic impulse. Gladwell does exactly that, blaming Rock's religiosity for his desire to make the Pill seem natural: "He was consumed by the natural. [...] In John Rock's mind the dictates of religion and the principles of science got mixed up."[9]

While Gladwell is right that the drive for a natural form of birth control is theological at heart, it's misleading to associate that theology solely with the dictates of the Catholic Church, or religious institutions more broadly. As we have seen, the worship of what's natural is common to secular and religious people alike, and the case of birth control is no different.

Long before Rock and the Pill there was fierce opposition to "artificial" forms of birth control from an ideologically diverse range of people. The nineteenth-century medical establishment, Catholics and otherwise, generally advised on secular grounds against using condoms and *coitus interruptus*, but approved periodic abstinence as "nature's own way" of preventing pregnancy. The scientific consensus that unnatural sexuality would lead to mental, physical, and moral ruin was echoed by traditional religious moralists and some of their staunchest opponents, including the earliest "free love" advocates. In an anti-marriage pamphlet titled *Cupid's Yokes*, the self-proclaimed Free Lover Ezra Heywood dismissed condoms and *coitus interruptus* as "unnatural" and "injurious." So, too, were celibacy and life-long monogamy: "The secret history of the human heart proves that it is capable of loving any number of times and persons, and that the more it loves the more it can love. This is the law of Nature, thrust out of sight and condemned by common consent, yet secretly known to all."[10]

Writing of Heywood and other liberal feminist reformers who despised birth control, the historian Linda Gordon argues that they "reflected a romantic yearning for the 'natural,' rather pastorally conceived."[11] Though their assessment of marriage (unnatural!) starkly contradicted that of religious contemporaries (natural!), it was nevertheless rationalized in the same way: by appealing to immutable laws built into the natural world, which man tampers with at his peril. Faith in naturalness, not the Christian Bible, was the lowest common denominator in discussions of proper sexuality.

To this day, secular critics of the Pill tacitly reinforce theological reverence for the natural and deep fear of the unnatural. Testifying before the US Senate in 1970, Johns Hopkins gynecologist Hugh J. Davis delivered a dire warning about the Pill's potential effects on humans. "The synthetic chemicals contained in the 'Pill' are wholly unnatural," he declared, "and this attribute applies in addition to the production process and to its performance as soon as it is introduced into the human body."[12] Despite being opposed to the Pill, Davis saw no problems with intrauterine devices, which did not contain synthetic chemicals and so, presumably, did not pose a risk to women's "natural" hormonal rhythms. Indeed, he invented one such device, the Dalkon Shield, marketed as a safe alternative to the unnatural Pill. By 1974 Davis's invention had become the nation's most popular IUD, prescribed to 2.2 million women.

But the Dalkon Shield was not safe. The device's nylon tail string, designed to be used for retrieval, also wicked bacteria from the vagina into the uterus, leading to miscarriages, hundreds of thousands of infections, and, in at least eighteen cases, death.[13] Legal woes followed and the device was eventually taken off the market, a reminder that safety, not naturalness, is the right criterion for legality.

More recently, the 2013 book *Sweetening the Pill*, by eco-feminist Holly Grigg-Spall, offers a critique of the Pill as a tool of capitalist patriarchy that undermines humanity's connection to the natural world and its rhythms. According to Grigg-Spall, women, along with all "oppressed races and classes," are "associated with concepts of 'Nature,'" and artificial forms of contraception are part of a general assault on Nature itself. Women naturally have a "strong connection" to the "lunar cycle," and despite John Rock's efforts, the Pill can only provide a fake version of that connection.[14] "The stream of synthetic hormones," writes Grigg-Spall, "does not function or fluctuate as the natural hormone cycle would." Like the Catholic Church, and like Dr. Hugh Davis, Grigg-Spall has a definition of natural that makes room for at least some permissible forms of contraception; in her case, NFP and condoms. Openness to condoms does not endear her to Catholics—one reviewer in a Catholic magazine complains that Grigg-Spall fails to see how "all contraceptive paraphernalia" is "an insult to the body."[15] This is, admittedly, a significant point of disagreement about what constitutes a violation of nature, but it points to an equally significant theological agreement about the importance of not violating nature in the first place.

Even Gladwell, who disparages Rock's theological devotion to nature, nevertheless endorses a different version of it. The title of his essay is "John Rock's Mistake," which at first appears to describe Rock's zealous pursuit of a natural pill. But Gladwell soon clarifies: Rock's real mistake isn't his idealization of naturalness, it's that he got nature wrong. Citing work by the scientist Beverly Strassmann, an expert on menstrual cycles in pre- and post-industrial women, Gladwell argues that regular monthly menstruation "is in evolutionary terms abnormal." In pre-industrial societies, onset of menarche occurs later in a girl's life and menstruation is far less frequent. The "lunar cycle" to which Grigg-Spall connects women is, in fact, unnatural. As always, unnatural is synonymous with "bad," and Gladwell suggests that numerous health problems, from mood disorders to breast cancer, may be due to the "unnatural" menstrual cycles caused by modern industrialized culture. The underlying assumption, again: we can debate the meaning of natural, but its goodness is indisputable.

There is, however, a brief aside in Gladwell's article that touches on a terrible inadequacy in the association of naturalness and goodness. After detailing the reasons for Dogon women's naturally infrequent menstruation—near-constant pregnancy and breastfeeding—he notes, in a parenthetical, that "those who survive early childhood typically live into their seventh or eighth decade." The same nature that protects Dogon women from diseases of modernity apparently has no qualms about executing their infants.

Needless to say, the Catholic Church does not forbid unnatural medical interventions that thwart naturally high levels of infant mortality. (This stands in contrast to certain extreme forms of nature worship like Christian Science, which rejects all medical technology in favor of "supremely natural" healing through God's will.) Nor, for that matter, does the Church forbid the unnatural breeding of plants and animals that forms the foundation of all agricultural food systems—modifications that allow for caloric excess, which, in turn, sets the stage for the wildly high levels of fertility that make some kind of birth control a virtual necessity, lest we become, to borrow from Pope Francis, "like rabbits." As Pope Paul VI observed, there are only "certain" limits beyond which it is wrong to go, only certain natural functions that must be preserved.

Well, which are they? Which natural functions are the ones that should be left alone, the ones that should be protected by law, enshrined in medical recommendations, sacred to God's plan? Which forms of contraception are approved by nature, and which should be forbidden?

The history of confused attempts to answer these questions is a case study in how appeals to nature are used to disguise sloppy logic, religious ideology, and visceral disgust.

Like all defenders of what's natural, Dr. Latz must concede the obvious in *The Rhythm*: certain forms of interference are entirely acceptable, whereas others are not. He picks haircuts and vomitoriums as illustrative contrasting examples, and in the course of two short paragraphs manages to reveal nearly every problem with this kind of argument:

> Not every interference with nature or natural processes is a serious matter. Cutting one's hair or finger nails is an "interference" with nature, but evidently of no consequence.
>
> The practice of the old Romans to eat their fill and then artificially to throw up what they had eaten and resume where they had left off at the banquet, is an interference with nature of a vastly different type. Such procedure is disgusting and means doing violence to nature.[16]

Latz's crude binary raises innumerable questions. Are tattoos a violation on the order of vomiting up your food, or are they more like cutting your nails? What about ballet, which can transform the shape of your feet? What criteria is he using? Tellingly, Latz does not mention health outcomes.

Instead, he simply calls the Roman practice "disgusting," which functions here as a synonym for doing violence to nature. (According to Church teachings tattoos are just fine, and some early Christians would even deliberately scar themselves with the wounds of Christ.)

With the criteria of disgust now foremost in the reader's mind, Latz turns to contraception:

> When we enter the innermost sanctuary of nature, that sphere of activities which is unique in its centrality, its profundity, its intimacy, its mysteriousness, its sacredness, its power for evil, we can readily see that the interference implied in contraception represents a violence to nature, a degradation, desecration, and perversion of the utmost seriousness [...] that it is in truth the *"crimen nefandum,"* "the unspeakable vice."[17]

Understood in the context of Latz's previous analogy, contraception is a disgusting perversion of sexual activity and therefore does violence to nature. It allows people to enjoy the pleasure of sex divorced from the purpose of procreation, just as the Romans enjoyed the pleasure of eating divorced from the purpose of nutrition.

But why would the rhythm method, employed to allow heterosexual couples the pleasure of sex divorced from the purpose of procreation, evade the vomitorium critique? Numerous influential figures thought it did not. In the nineteenth and early twentieth centuries, both the possibility and the actuality of having children were thought to act as nature's check on rampant lust. As veteran radio preacher Father Ignatius Cox put it in a 1930 broadcast, "Children thus check and control sex."[18] Contraceptives, like anal and oral intercourse, were associated with prostitutes, who needed to avoid the natural consequences of their professional activities. Moralists like Cox saw the rhythm method as yet another potential tool of blasphemous Free Lovers who sought to divorce sex from the proper end of procreation. Timing sex to avoid fertile periods would "pervert the order of nature," declared Monsignor Louis Nau, a staunch opponent of natural birth control—for him, a deceptive misnomer—just two years after Latz first published his bestseller in 1932.[19] According to Father Charles Coughlin, another enormously popular radio preacher who was eventually forced off the air for anti-Semitism and fascist sympathies, non-procreative marital intercourse was nothing more than "legalized prostitution."[20]

Latz obviously disagreed, but aspects of his counterargument were pe-
culiarly circular. "There is nothing reprehensible in enjoying the pleasures
of sex without having to bear the burdens ordinarily resulting therefrom," he
maintained, "provided no violence is done to nature, no law of God is violated."[21] It's a textbook case of begging the question. Does the rhythm method
do violence to nature? No, because it does not pervert the order of nature.
What does it mean to pervert the order of nature? Doing violence to nature.

To escape the circularity it's helpful to consider Latz's characterization
of contraception as a *crimen nefandum*. This legal term of art, literally "the
unspeakable crime," was typically understood as referring to sodomy, which,
as the eighteenth-century British jurist William Blackstone asserted, was "a
subject the very mention of which is a disgrace to human nature."[22] Sodomy
was the quintessential "crime against nature" (*contra natura*), another important legal term of art used to describe multiple sexual activities, including
masturbation, bestiality, incest, oral intercourse, and anal intercourse—any
sexual act that did not lead to procreation. For medieval scholastics, sodomy
was the worst of all crimes against nature, outranking rape and incest in
the hierarchy of sins. Legal punishments were meted out accordingly, with
sodomites executed alongside murderers, and heterosexual rapists getting
off with lighter sentences.

Since contraception frustrated the natural purpose of sex, it too was
classed as a grave crime against nature. St. Augustine wrote ferocious attacks
against intercourse undertaken for any reason other than begetting children,
employing the same analogy to food that Latz did: "What food is to the
health of a man, intercourse is for the health of the species." Contraception
was seen as akin to abortion, and, therefore, a kind of murder: "As often as
she could have conceived or given birth, of that many homicides she will
be held guilty," argued an influential sixth-century French bishop, taking a
position that remained common for centuries.[23]

There are several reasons such crimes were considered "unspeakable,"
when gruesome crimes like murder were not. One was general prudishness
about all sexual activity. (Augustine called sexual organs "pudenda," which
means "objects of shame.") Another reason was to hide the existence of such
shameful crimes. According to the historian Theo van der Meer, up through
the eighteenth-century, executions of homosexuals occurred in secret and
"corpses of such culprits had been thrown into the sea and buried under
the gallows," the better to forget them.[24] In addition to preserving the good

name of communities and nations, secrecy ensured that people wouldn't hear about sodomy to begin with, which could inspire more unnatural behavior. Censorship, it was hoped, would stop the spread of vice.

In the United States, that same censorial attitude led to an 1873 federal act and subsequent state laws known as Comstock Laws, which criminalized the distribution by mail of any "vile, filthy, or indecent thing." They were named for their chief advocate, Anthony Comstock, a fervent moral reformer who boasted of having confiscated 160 tons of obscene literature and 64,094 "articles for immoral use, of rubber, etc.," and, more darkly, of having caused fifteen deaths by suicide of publishers, dealers, and manufacturers of obscene material.[25] Comstock adopted an expansive notion of obscenity, and so did his eponymous laws, which covered sex toys, pornography, and even descriptions of sexual activity. This last led to the confiscation of anatomy textbooks with pictures of human beings' shameful parts. (Ezra Heywood, the Free Lover and author of *Cupid's Yokes*, was arrested by Comstock in 1877 for mailing his pamphlet and then imprisoned for two years.)

All forbidden material was considered dangerous for the same reason: it could lead to crimes against nature. The laws singled out the worst of such crimes, contraception and abortion, for special attention, prohibiting "every article, instrument, substance, drug, medicine, or thing which is advertised or described in a manner calculated to lead another to use or apply it for preventing conception or producing abortion, or for any indecent or immoral purpose." In effect, the laws criminalized contraception, since there was no way to distribute contraceptives or information about them by mail. (As of this writing, crime against nature laws remain on the books in nine US states, though the constitutionality of their enforcement has been called into question. Massachusetts law states: "Whoever commits the abominable and detestable crime against nature, either with mankind or with a beast, shall be punished by imprisonment in the state prison for not more than twenty years.")[26]

Concern about crimes against nature was not a minority view held by religious zealots. A public inculcated with Victorian morality had grave misgivings about shifting attitudes toward sexuality, facilitated by new technology that allowed for better production, distribution, and advertisement of "indecent" things. "The anti-contraceptive laws were not originally passed as a result of controversy over religious doctrine," emphasizes the historian Carol Flora Brooks. "They were passed as a by-product of an attempt to give legal support to a widespread attitude about obscenity. Virtually the only opposition to their passage came from the fear of a small minority for the freedom of the press."[27]

The advent of NFP allowed moralists, religious and secular alike, to meet the demand for effective contraception while simultaneously maintaining allegiance to their traditional, natural moral order. The compromise between morality and practicality was very appealing and easy to rationalize. After all, early Church fathers defended sex during pregnancy and after childbearing age, both of which were necessarily "unfruitful." If sexual intercourse without the purpose of procreation was permissible in circumstances when nature had made procreation impossible, then surely there was room for another natural exception.

However, accommodating NFP demanded a redefinition of "natural" intercourse. No longer could it mean "intercourse directed towards procreation." Instead, "natural" intercourse came to mean the depositing of semen into the vagina in the context of marriage. This definition preserved the traditional moral order that prohibited masturbation, oral and anal intercourse, and other forms of contraception associated with prostitutes. At the same time, it allowed married couples of all ages to enjoy the pleasure of regular sexual intercourse without the consequence of procreation. Theologians shifted away from emphasizing procreation as the primary end of sex, choosing to foreground the sexual act's natural tendency to bring married couples closer to each other. Pope Pius XII clarified for the public: the "observance of the sterile period can be licit" for any couple with "serious reasons," which included, quite broadly, "social" or "economic" motives.[28]

Where did the Pill fit into this redefinition of natural intercourse? The answer was far from clear. Pope Paul VI himself expressed confusion, stating in 1965 that he did not know the answer to the problem of artificial contraception. (He ended up prohibiting it in 1968.) His friend, the influential Catholic philosopher Jacques Maritain, thought it should be allowed. "In my opinion there is no essential difference between the Pill and the other methods, between a mental calculation and a medical intervention," argued Maritain. "Their moral condemnation cannot be justified by reason." Certain uses of the Pill, he reasoned, could be "in conformity with nature."[29]

Maritain saw that opponents of artificial contraception would be forced to take increasingly tenuous positions on the meaning and morality of natural sexual activity. If "natural" means the eventual depositing of semen into

a vagina, then infertile couples can still have natural sex, but a couple incapable of vaginal intercourse is forbidden from any sexual activity. If a man
becomes impotent later in marriage, his wife can never again receive sexual
pleasure. A twenty-year-old war veteran whose genitals are mutilated during
combat cannot marry his sweetheart when he returns, because they would
not be able to consecrate their marriage with natural sex.

These are not random examples. They are current Church doctrine. I
actually borrowed the example of the war veteran from a 2014 article in a
Catholic magazine that defends the prohibition on his getting married. "A
critic might say [...] the Church should marry an impotent couple so that
they can licitly engage in sexual activity they are physically able to enjoy such
as passionate kissing, fondling, mutual masturbation and oral stimulation,"
writes Trent Horn.[30] But, on his account, the hypothetical critic, though
well intentioned, fails to understand that the veteran's marriage would violate natural law. Sexual activity *must* end in penile-vaginal intercourse. Everything else is a means to that end. Horn makes his case with the exact same
eating metaphor employed by Latz nearly a century earlier, and Augustine
centuries before that: "Reducing sex to only these activities is like reducing
eating to only chewing and tasting food without digesting it."

This definition of natural sexuality also opens up a range of allowable activities that strain any reasonable understanding of natural, however defined.
Dildos and vibrators? Like the rhythm method, they are completely permissible according to Gregory Popcak, author of *Holy Sex: A Catholic Guide
to Toe-Curling, Mind-Blowing Infallible Loving*, endorsed by a raft of leading
Catholic intellectuals. All you have to do is follow what he calls "the one
rule," namely the consummation of sexual activity by ejaculating inside of
the vagina. (In a meaty footnote, Popcak explains that even anal intercourse
with a condom is not technically forbidden as a part of foreplay, provided it
leads to vaginal intercourse.)[31]

Despite approving of deliberate celibacy during periods of natural fertility, and despite allowing for artificial forms of stimulation, Popcak leans
heavily on the wisdom of Mother Nature throughout his book. In this he
resembles secular critics of the Pill such as Grigg-Spall, who also stresses
the danger of unnaturally violating biological systems—and the benefits of
aligning with them. "The presence of semen in the vagina causes a powerful chemical chain reaction in the woman, and the absence of this chemical
reaction cannot help but impede the physiological process of attachment,"

Popcak tells his readers. "When the Church teaches that barrier methods of contraception (as well as other forms of artificial birth control) offend the uniting power of sex, the Church may be speaking more literally—and, considering our emerging understanding of interpersonal neurobiological processes, more profoundly—than most people think."[32]

The weaknesses in these modern appeals to nature are readily apparent. Using sophisticated apps, thermometers, and cutting-edge scientific knowledge to time intercourse precisely when it *won't* result in procreation ("sympto-thermal" birth control) seems like the very definition of artificially thwarting nature's intentions, even if the company selling the app is called "Natural Cycles." And there's something suspicious about the neat complementarity of Popcak's biologically based arguments with his preferred approach to sex. Why would condoms and *coitus interruptus* have serious biological consequences because they violate nature's design, but deliberately refraining from intercourse during natural periods of fertility is completely fine? (That's not to mention clerical celibacy, which, since it demands *never* using one's genitals as nature intended, should lead to all kinds of biological and spiritual problems.)

Not only that, but far from naturally bringing all couples together, NFP often poses grave intimacy problems. In a report delivered to the Pontifical Commission on Birth Control in 1965, commissioned by the international Christian Family Movement, devout Catholics confessed deep doubts about the rhythm method. "Rhythm destroys the meaning of the sex act," wrote one husband. "It turns it from a spontaneous expression of spiritual and physical love into a mere bodily sexual relief; it makes me obsessed with sex throughout the month; it seriously endangers my chastity; it has a noticeable effect upon my disposition toward my wife and children." His conclusion? "Rhythm seems to be immoral and deeply unnatural. It seems to me diabolical."[33] Biological evidence to support his position was provided in the form of research suggesting that female sexual desire is naturally strongest during times of fertility.

Faced with these arguments, commission members began to waver, and in the end a strong majority voted against the view that contraception is intrinsically evil. (The current prohibition as articulated in *Humanae vitae* is the result of a minority theological opinion, pushed through by a series of high-level political maneuvers.) In the words of Pulitzer Prize–winning historian of religion Garry Wills, himself a Catholic, the members were forced to "look honestly at the 'natural law' arguments against contraception and see, with a shock, what flimsy reasoning they had accepted."[34] Wills goes on to attack the food analogy. "Eating is for subsistence. But any food or drink beyond what is necessary for sheer subsistence is not considered mortally sinful. In fact, to reduce eating to that animal compulsion would deny symbolic and spiritual meanings in shared meals—the birthday party, the champagne victory dinner, the wine at Cana, the Eucharist itself."[35]

Wills suggests that in the context of sexuality, appeals to nature really serve to disguise the disgust that motivates them. A generalized "fear and hatred of sex" lies behind the logic of crimes against nature, whether those crimes are in the form of sodomy or contraception. It is for this reason that Latz foregrounds the disgustingness of Roman vomitoriums—his argument against unnatural sexuality, like all such arguments, depends on provoking disgust in his readers. The logic, science, and evidence adduced by advocates of natural sexuality do not lead to a dispassionate conclusion; rather, they function as post hoc justifications of theologically inflected disgust.[36]

In a legal review article on "crimes against nature," drafted just after the Kinsey Reports on sexual behavior exploded popular perceptions of normal sexuality, the author expresses concern about the distorting force of visceral disgust: "One of the basic difficulties in setting up standards on which to draft statutes for the control of sexual deviations comes from the fact that society condemns them so severely because of their repugnance to moral conventions, rather than because of any actual physical harm caused by them."[37]

Perhaps more than any other sin, more than thievery, fraud, or even murder, unnatural sex—however a given ideology happens to define it—generates visceral disgust and outrage. Politicians convicted of embezzling can win elections. A politician convicted of bestiality? Inconceivable—which is strange, given that one crime seems more obviously related to political duties than the other.

Legislating according to an ad hoc combination of convention and repugnance is dangerous. And when it comes to sexuality, it is wise to be suspicious of appeals to nature, which are too often smuggled in to cover for reflexive loathing or religious preferences. Otherwise, we risk repeating the same kinds of mistakes that stain our legal past.

Deep-seated terror of unnatural sexuality can turn humans into monsters, as it did to those Americans who participated enthusiastically in lynchings.

Accounts of lynchings are unimaginably horrific, and the 1899 murder of twenty-four-year-old Samuel Wilkes can be taken as representative. It was a spring afternoon in Georgia and his murderers were lusting for justice. Their victim had been accused of killing his white employer, Alfred Cranford, in Newnan, a rural town just outside of Atlanta. But the mob didn't form until rumors began circulating that Wilkes had also assaulted Mattie Cranford, Alfred's wife. The alleged violation of Mattie Cranford was seen as a violation of the natural order, and it sent white residents of the small town, at least five hundred of them, into a frenzy of evil. Before tying Wilkes to a tree and burning him alive, the mob sliced off his ears, removed several fingers, and castrated him. The flames had yet to die down when they descended like vultures onto Wilkes's charred corpse. "Not even the bones of the Negro were left in the place," wrote the activist and investigative journalist Ida B. Wells. "[They] were eagerly snatched by a crowd of people drawn here from all directions, who almost fought over the burning body of the man, carving it with knives and seeking souvenirs of the occurrence."[38]

Along with the brutality, the accusation of sexual assault and subsequent castration were characteristic of American lynching. An exhaustive 2017 study conducted by the Equal Rights Initiative determined that fear of interracial sex was the "most consistent and common provocation that could create mob violence directed at African Americans."[39] The same fear motivated laws against interracial marriage that existed in twenty-five states at the time of Wilkes's murder. Essential to these legal prohibitions was the idea that sexual relationships between races went against nature, and therefore against God. Interracial sex was likened to bestiality—"this slavery of white women to black beasts," as Georgia congressman Seaborn Roddenberry described it, just a decade after his state hosted Wilkes's lynching.[40]

In her study of nineteenth- and twentieth-century miscegenation law, *What Comes Naturally*, Peggy Pascoe shows that "on the unnaturality of miscegenation judges were a virtual chorus." Perhaps the most influential decision was written in 1865 by a Pennsylvania judge in support of segregated railcars:

> God has made [the races] dissimilar, with those natural instincts and feelings which He always imparts to his creatures, when he intends that they shall not overstep the natural boundaries he has assigned to them. The natural law which forbids their intermarriage and that social amalgamation which

leads to a corruption of races, is as clearly divine as that which imparted to
them different natures. . . . The natural separation of the races is therefore
an undeniable fact.[41]

Egregious violations of natural law and divine law were thought to have
disastrous civic consequences. Like polygamy, sodomy, adultery, and incest,
interracial sex was seen as an unnatural threat to the civic institution of marriage, upon which social harmony depends. Pascoe collects numerous examples of how this assumption served as a foundational justification in legal
reasoning. In 1877, the Alabama Supreme Court warned of the "discord,
shame, [and] disruption" that interracial marriage "must naturally cause."
A year later, the Virginia Supreme Court invoked the specter of civiliza-
tional collapse if "alliances so unnatural that God and nature seem to forbid
them" were allowed to flourish. "The amalgamation of the races," opined
a Georgia court in 1869, "is not only unnatural but is always productive of
deplorable results."[42]

To decisively defeat antimiscegenation laws, activists and intellectuals
focused on changing the public's understanding of natural law. The idea of
"natural aversion," argued George Schuyler, a prominent African American
author and newspaper editor, and himself a member of an interracial marriage, was entirely invented. "On the contrary, the researches of scientists
prove that there is a great deal of attraction between them."[43] His friend,
the journalist J. A. Rogers, focused on the anthropology of sexual attraction
in a series of books—*As Nature Leads, Sex and Race, Nature Knows No Color-
Line*—that explicitly appealed to nature as authorizing interracial marriage.
"Sex relations between so-called whites and Blacks go back to prehistorical
times," he wrote. "Races have always mixed, and will continue to mix; if
not in accordance with man-made laws, then against them."[44] God had not
intentionally separated the races, and neither had nature. (As evidence, they
could cite statements from the Vatican itself, which supported the valid-
ity of interracial marriage and condemned in no uncertain terms the notion that African Americans constituted a biologically inferior branch of
humanity.)[45]

If racial mixing was a "perfectly natural process," as Rogers believed,
then interracial marriage was a fulfillment of natural law, and outlawing
it was unnatural. After chronicling the eventual political and legal success
of this position, Pascoe reminds us that it depended on other assumptions

about what nature "wants," assumptions with unforeseen consequences. Heterosexual desire became the primary defining characteristic of natural sexuality and marriage, with obvious implications. "In this respect," writes Pascoe, "Schuyler and Rogers played a role in producing a modern culture that increasingly assigned its fears of unnaturality to homosexuality rather than to race mixture."[46]

To this day, opponents of same-sex sexual activity and same-sex marriage routinely reference perversions of nature and the civic danger posed by unnatural unions, in language that echoes earlier arguments about the danger of interracial sex and marriage. The Republican Party's official 2016 platform adopted a definition of "natural marriage" as the "union of one man and one woman," a relationship that is the "cornerstone of family" and the "foundation of civil society." Opponents of interracial marriage got nature wrong, yes, but opponents of same-sex marriage are getting nature right. Despite mentioning that families depend on God, the authors of the platform claim their position does not depend on "ideology or doctrine." It is secular because it is "natural," fully discoverable by "social science and common sense"—though, of course, what's natural is also approved by God.[47]

Ironically, defenses of same-sex relationships echo earlier arguments about the naturalness of interracial sex. The historical ubiquity of same-sex attraction; the existence of homosexuality in nonhuman animals; evidence of evolutionary selection for homosexuality: all of these are taken to demonstrate that same-sex attraction and relationships are part of the natural order, and legislating against them is a violation of that order. The biological basis of sexual orientation is stressed by LGBT advocates and organizations like Planned Parenthood: "Sexual orientation is a natural part of who you are—it's not a choice."[48]

And if same-sex desire is natural, might it not mean that God approves, too? When God and Nature are convertible, then it must. As Lady Gaga sings in her anthem about LGBTQ tolerance, if people are "born this way" it can't be wrong, "'cause God makes no mistakes."

The naturalness argument might be at its most persuasive when it comes to sexuality. For decades, popular opposition to homosexuality depended on its supposed absence in the animal kingdom. When scientists published a 1977 study documenting homosexual behavior in seagulls, public outrage from anti-gay groups was swift and severe, and so was celebration from gay rights advocates.[49] Today the naturalness of homosexual behavior in animals and humans is well established, and this is taken by some, like Republican senator Orrin Hatch of Utah, as proof of its inherent morality. "I don't think people choose to be gay," said Hatch in 2018. "I think they are born that way and if they don't choose to be gay, then how do you blame them for it?"[50]

Other attempts to break down sexual taboos make use of appeals to the natural world and the natural state of the human species. The bestseller *Sex at Dawn* defends alternatives to monogamy by surveying the prevalence of nonmonogamous sexuality in preagricultural societies as well as our primate relatives. "We *are* apes," declare the authors. "Like bonobos and chimps we are the randy descendants of hypersexual ancestors."[51] The standard story of natural monogamy is wrong, goes the argument, which means the standard view of monogamy's morality might also be wrong.

This is the wrong way to think about sex and morality. The real problem is not that we got natural sexuality wrong for so long. It's that moralizing and legislating about sexuality on the basis of its perceived naturalness is a mistake. Forget the seagulls. There's a significant scientific consensus that although homosexual behavior occurs in the natural world, same-sex coupling is exceedingly rare. When it happens in birds, for example, scientists believe it is because of an absence of males, not an intrinsic biological sexual orientation.[52] In a world where naturalness determines morality and the laws that enforce it, the new science would be a devastating blow to LGBT rights. But the sexuality of birds—or primates, or preagricultural humans, or whatever version of nature we choose—has precisely nothing to do with whether a religion or a nation should prohibit same-sex relationships.

In the case of a religion, appeals to scripture might be appropriate. Calling those scriptural teachings natural, however, is sleight of hand, an attempt to backfit theological laws into an improvised version of naturalness that can justify them. And in the case of a non-theocratic nation, neither the scriptures of a religion nor the false god of nature should dictate people's sexuality. Instead, we should depend on secular criteria such as the reduction of suffering or our rights to liberty and the pursuit of happiness—which, contra Jefferson, are not guaranteed by Nature's God.

Calling a sexual practice unnatural is an easy way to avoid revealing the true reasons for a law against it, be they disgust, religious devotion, or personal preference. Bracketing theological appeals to nature forces people to be clearer about the foundations of their beliefs, and the laws that follow from them.

However, bracketing appeals to nature does not entail dismissing the value of naturalness entirely, or the utility of nature as a regulatory standard. National parks, for instance, cannot disregard the value of naturalness, since their identity is built around the idea of preserving nature. Unfortunately, the tendency of naturalness to inspire extreme positions makes it difficult to find a middle ground between worshiping nature and ignoring it. In the next chapter we will see how regulatory debates in sports center on the difficulty of demoting nature from divine status without dismissing the value of naturalness entirely.

CHAPTER 9

God-Given Talent

"CAN YOU SMELL IT?" asked Marjorie Thrash, an accomplished body-builder who took a break from her administrative duties to give me a back-stage tour at the Organization of Competition Bodies (OCB) Atlantic Super Show Pro/Am in Richmond, Virginia.

It's a rhetorical question. We're surrounded by bodybuilders in various states of undress, and the sharp scent of chemical tanning solutions hangs heavy in the air, impossible to miss. The competitors look like plastic action figures come to life, Oompa-Loompa orange and lacquered ebony, muscles swollen and gleaming, their skin latex-thin thanks to dietary restrictions and "sodium cycling," a pattern of loading and lowering sodium intake that de-pletes subcutaneous water. "When you hit that stage you're not in a natural state," explained Thrash of the grueling routine that precedes a competition. "A week later you don't look like that anymore. It isn't natural to be that lean."

Between the extreme leanness, the alien skin tones, the strained poses, and the continual smiling, it's hard to imagine any human form more delib-erately artificial than these sculpted athletes. And then there's the artificiality of the training: regimented workouts, specially designed equipment, scru-pulously monitored diets, the protein powder and the creatine supplements in giant plastic tubs with names like Syntha-6 and ENGN Shred. Some are being advertised in booths outside the Atlantic Super Show auditorium, each promising a unique competitive edge to bodybuilding hopefuls. The branding evokes a pharmacy, not a forest, with technical specifications and scientific guarantees overlaid on pictures of metallic molecules. Nature, transcended with technology.

Yet this particular show, like all OCB shows, is distinguished by the fact that it is only open to natural bodybuilders. Every entrant must pass a drug test and a polygraph to ensure they are not taking any banned substances. These include: anabolic agents (the OCB website lists forty-five separate types), anti-estrogens, peptide hormones, growth hormone–releasing hormones, selective androgen receptor modulators, prescription-grade diuretics, prescription-grade stimulants, prescription-grade weight-loss substances, and synthol, a "site-enhancement oil" that's injected to "fluff" lagging muscles, sort of like Botox for bodybuilders.

By contrast, the International Federation of Bodybuilding and Fitness, which administers the Mr. Olympia competition, is known for deliberately lax standards when it comes to performance-enhancing drugs. Though they claim to administer random drug tests—and top athletes claim never to use drugs—everyone in the bodybuilding world knows the truth. If you choose to go "natty" you'll never have the physique of Mr. Olympia champions, whose massive muscles make the competitors I saw in Richmond look positively scrawny. Online bodybuilding forums like NattyOrNot.com routinely mock those naïve enough to believe otherwise. "Is it possible to build a Ronnie Coleman body naturally without using steroids?" asks one user on the popular question-and-answer website Quora. Ronnie Coleman is an eight-time Mr. Olympia winner, widely regarded as one of the greatest bodybuilders of all times. The responses, from natural and drug-using bodybuilders alike, are unanimous. "Absolutely *no way*," declares the top-rated reply. "Ronnie is a freak of nature, with incredible hulk genetics. . . . Even with massive amounts of steroids, it's extremely unlikely."[1]

Freak of nature. This common phrase highlights a fundamental paradox that arises when defining naturalness. The second entry for "natural" in the *Oxford English Dictionary* is "consistent with nature; normal, ordinary." According to this definition, freaks of nature are unnatural because they are abnormal and extraordinary. By logical extension, world-class athletes are all unnatural. "Freak of nature" applies not only to Ronnie Coleman, but to every athlete blessed with the genetic gifts necessary to become the best in the world at a given sport. (Phil Heath, seven-time Mr. Olympia, is nicknamed "The Gift.") Nor is it just their biology. The willpower and lifestyle required to convert freakish genetics into world-class athletic performance are also, in that sense, unnatural. Normal people do not log thousands of hours hitting a ball with a stick or throwing a ball through a hoop. Even excelling in more natural-seeming sports such as running requires unnatural methods. The fastest hunter-gatherers in history would fail to qualify for the Olympics at any distance, because doing what comes naturally just isn't enough to be among the fastest people in the world.

At the same time, the "of nature" that comes after "freak" acknowledges that deviations from the norm can be natural. Shaquille O'Neal's abnormal height is completely natural, as are Michael Phelps's abnormally long arms and torso. So, too, is primary polycythemia, a genetic mutation that raises your red blood cell count, which helped the Finnish cross-country skier Eero Antero Mäntyranta collect seven Olympic medals in the 1960s. These deviations are "of nature" in a way that height gained by using stilts, or legs lengthened with flippers, or red blood cell count raised by injecting extra cells, are not. Elite athletes are all naturally unnatural—even the ones who take banned substances. "Definitely no judgment about the guys who are juicing," said Lenwood Hall, a forty-one-year-old bodybuilder with streaks of gray in his beard who spoke with me at the Richmond show. "It's like when everyone said Barry Bonds hit all those home runs because of drugs. I think, 'Naw, you could juice and you still wouldn't be able to do that.'"

All of this makes defining naturalness in the context of sports very difficult. When I asked Hall what natural meant to him, he told me it was mainly about health and safety. "This stuff is tearing up livers and kidneys," he said. "The guys who take it, they look rough, they look ten years older." Thrash echoed his emphasis. "We've seen a number of athletes die," she said. "People in natural bodybuilding want to be healthy."

Their perspective is borne out by the regulations in natural bodybuilding. Testosterone and growth hormones are banned; breast implants are not—presumably because, in theory, the former is more dangerous than the latter. It certainly can't be any intuitive understanding of naturalness, given that the body naturally produces testosterone and growth hormones, not silicone gel. Nor is it that one confers an advantage and the other doesn't: experts estimate that over 80 percent of female bodybuilders have implants, which are helpful for pleasing judges, especially in the "figure" division where feminine form is one of the criteria.

Despite these apparent contradictions, the distinction between natural and unnatural athletes persists, often with religious overtones. Athletes who compete without drugs are referred to as "clean": morally, insofar as they haven't cheated; but also physically, as their bodies are unpolluted by substances that, according to the laws of sporting organizations, could transform their performances into artificially enhanced frauds. "You can do this clean and with just God given talent," said Maurice Greene, a former world record holder in the 100-meter dash, in 2005.[2] Just a year earlier, Wisconsin representative Jim Sensenbrenner used the same explicitly religious language to justify the Anabolic Steroid Control Act before Congress: "We should admire the athletes who achieve greatness through hard work and their own God-given abilities."[3]

If natural just means "healthy," then one way to resolve the paradox inherent in talking about natural athletes is doing away with the idea of naturalness altogether. Instead of natural bodybuilders, call them safe bodybuilders. "Safe" might still pose definitional and regulatory difficulties—you'd have to define and test for "unsafe" levels of amphetamines or anabolic steroids—but at least it means what it's supposed to mean, and doesn't conflict with the obvious artificiality of sports as a practice. This move could even lead to the legalization of some banned substances such as growth hormones, prohibited because of an irrational bias against the perceived unnaturalness of using them. As one article argued in the *British Journal of Sports Medicine*, "drugs that improve our natural potential" could be a welcome part of athletic activity, provided they are safe. "Far from being against the spirit of sport," maintain the authors, "biological manipulation embodies the human spirit—the capacity to improve ourselves on the basis of reason and judgment. When we exercise our reason, we do what only humans do."[4]

Emphasizing safety and deemphasizing naturalness help to avoid incoherent arguments about the meaning of performance-enhancing drugs and God-given abilities. Take this endorsement in *Golf Digest* of cortisone, a synthetic corticosteroid that athletes may use legally: "Cortisone is a synthetic steroid designed to mimic the natural version the body produces in low doses to help heal injuries. It's not an anabolic steroid, so it doesn't have performance-enhancing properties beyond making you recover faster than the normal healing process."[5] Why, one wonders, does "making you recover faster than normal" not qualify as a type of performance-enhancement that ought to be banned? And why does it matter that cortisone mimics something produced in the body? Blood-doping, which is illegal, involves no unnatural substances (aside from the equipment). Just remove some of your blood, wait for your blood volume to return to normal, then add in the blood you removed. As for God-given ability, it's hard to imagine how Sensenbrenner would be able to explain why synthetic cortisol injections honor it, but anabolic steroids do not. Shouldn't everyone just rely on their natural (read: normal) recovery processes? God-given abilities and their natural analogue seem entirely irrelevant.

When I attended the show in Richmond, I was hoping to gather evidence in support of this perspective. The absurdity of "natural" bodybuilding—the stark contradiction between the assertion of naturalness and the unnatural-ness of the athletes—could be exported to other sports. If "natural" isn't syn-onymous with "good," then there's no need to be "natty," no basis to claim that some athletes are honoring their God-given talents and other athletes aren't. I could show how, absent a religious idea of nature at the center of everything, the laws that govern sport could be made more consistent and inclusive by ignoring naturalness. We do not disdain mathematical proofs if the mathematician was found to be using performance-enhancing drugs. Imagine subjecting artists, musicians, and authors to drug testing. Sports, I thought, should be regulated in the same way we regulate other human displays of talent—with an eye to the quality of the product, not whether the producer took advantage of some artificial assistance.

But as I marshalled my arguments, I realized I was wrong. Key features of sports are hard to make sense of without some baseline ideal of natural athleticism, just as national parks make no sense without some baseline ideal of nature. Sports and national parks have multiple defining characteristics, and one of them is that they celebrate nature. The former celebrates nature as it occurs in human physiology, the latter as it occurs in ecosystems. In both cases the ideal is necessarily and justifiably compromised. It exists in tension with other ideals that are also extremely important—accessibility, for example. Yet ignoring the significance of nature altogether is no less a mistake than worshiping it—and the regulations that would result are equally problematic.

On August 30, 1904, with the devastating St. Louis summer sun beating down, an American bricklayer named Fred Lorz defeated thirty-one other men to claim victory in the third Olympic marathon.

Except Lorz hadn't actually won. After collapsing from exhaustion only nine miles in, Lorz hitched a ride in his manager's car for the next eleven miles, then hopped out and crossed the finish line. Cheers erupted, and Alice Roosevelt, the president's daughter, crowned him with a laurel wreath. But before she could add the gold medal, spectators revealed the fraud. Lorz fessed up and he was disqualified. The medal eventually went to Thomas Hicks, another American, who had nearly killed himself with a performance-enhancing cocktail of strychnine and eggs, and ungodly amounts of brandy (his manager ran out by the end), all of which would have disqualified him under current regulations.[6]

Strange as it may seem, Lorz is not the only person to cheat with a vehicle in a high-profile race. The Cuban American runner Rosie Ruiz was found to have taken the subway during the 1979 New York Marathon, qualifying her for the 1980 Boston Marathon—which she then won, stunning observers by sailing across the finish line without breaking a sweat. (It turned out she had jumped onto the Boston course with less than a mile to go.) And in 2016, the much-beloved British ultra-distance runner Robert Young, aka "Marathon Man," set out to break the record for running across America, only to be disgraced when investigators used satellite tracking data to determine that "long stretches of the attempt can only be explained by Young having ridden in or on a vehicle."[7]

Technically speaking, riding on subway trains and cars violates Rule 144.3(d) of the International Association of Athletics Federations' (IAAF) official competition rules, which prohibit

> the use of any mechanical aid, unless the athlete can establish on the balance of probabilities that the use of an aid would not provide him with an overall competitive advantage over an athlete not using such aid.[8]

A previous version of this rule gave some examples: "Any technical device that incorporates springs, wheels or any other element that provides the user with an advantage over another athlete not using such a device."[9] Exceptions are made for equipment specific to any given sport: shoes, in the case of running, a pole for the pole vault, and so on.

The idea here is both obvious and philosophically fraught: sports are activities performed by human beings using a highly regulated and standardized set of tools. "Mechanical aid" and "technical device" are the equipment equivalents of performance-enhancing drugs, artificially supplementing the physiological performance of the human being.

This division between *human being* and *equipment* is significant because it helps us to locate the source of an athlete's performance. That's why, when the Kenyan phenomenon Eliud Kipchoge ran a marathon in under two hours for the first time in history, a slew of articles made sure to note he was wearing an exclusive edition of Nike's Vaporfly shoes. Studies have confirmed Nike's claim that their high-tech foam and carbon-fiber plate improve marathon times by at least 4 percent. That information is crucial. It helps us to separate the role of natural talent perfected by training—the results of which are located in the athlete's mind and body—from the role of mechanical aid, which is not. Moreover, our knowledge of these respective roles affects how we feel about an athletic performance, and rightfully so. If someone comes along and smashes Kipchoge's record, the significance of their accomplishment can't be fully processed without knowing how much of a role was played by their shoes. It *matters* whether they did it barefoot or wearing some newfangled Vaporfly ULTRA—even if the new shoes are permitted under the rules.

These are not the kinds of questions we ask of an engineer who has designed an innovative and highly efficient engine. The task of designing an engine has nothing to do with showcasing the engineer's natural talent, whereas the task of running a marathon is meant to showcase an athlete's natural talent. It would be strange to feel disappointed upon hearing that the engine was designed using new technology, because that's not what we care about when it comes to designing engines. (On the other hand, it would be disappointing to hear that the engineer *stole* the basis for the design from someone else, because the value of respecting intellectual property is one that exists in the world of engineering.)

For a parallel example, consider the game of chess. To "play chess" really means "to be a human playing chess." Playing chess is an entirely different activity than creating software that generates successful chess moves. The software engineers behind the world's greatest chess engines would be crushed in a standard match against the world's reigning chess champion, just as the world chess champion would be crushed by the chess engines they designed. And if it were discovered that Magnus Carlsen, the world chess champion as of this writing, had in fact been aided by a supercomputer in the championship match, we would conclude that he hadn't really won—the talent was located in the computer, not in him.

The very existence of human chess champions as we know them depends on maintaining a distinction between biological humans and their artificial equipment. Since machines are much better at chess than humans, allowing humans to use whatever equipment they want would effectively turn chess competitions into software engineering competitions. We know what these would look like because they already exist in the form of computer chess tournaments. Such tournaments do not celebrate the ability of humans to play chess—they celebrate the ability of humans to design programs that generate excellent chess moves.

Superior shoes do not give the same advantage to runners that a chess engine gives to chess players, but the principle behind regulating them and the reason we care about their contribution to outcomes remain the same. At a certain point the role of mechanical aid overshadows the role of human performance, and the essence of the activity changes. The line where that happens, like many dividing lines, is a subjective moving target, but doing away with it entirely isn't an option—not, at least, for those who want sports to remain sports and not engineering competitions.

Rule 144.3(d) also governs the use of prosthetic limbs, a controversial regulatory arena where some standard of naturalness is indispensable.[10] When double-amputee sprinter (and eventual convicted murderer) Oscar Pistorius was initially disqualified from the 2008 Olympics, the IAAF's official statement cited research that showed Pistorius's prosthetic "Cheetah" blades—named, of course, for a biologically gifted animal—needed "less additional energy than running with natural limbs" and exhibited unique biomechanical characteristics that placed athletes with "natural legs" at a disadvantage. Pistorius appealed the decision and won, on the basis that the evidence was insufficient. In 2016, similar criteria were used to disqualify Markus Rehm, a single-amputee long jumper who had hoped to compete in the Rio Olympics.

In the context of the IAAF's statement, "natural" limb means "biological" limb, since that is the only conceivable comparative class for determining whether a prosthesis gives an unfair advantage. Unfair compared to what? A hypothetical biological limb. The scientists who collect the evidence informing these decisions use an array of sophisticated techniques to make the comparison. They cover study participants—with and without prosthetics—in reflective markers and have them run past a battery of infrared cameras. Force plates measure the downward force exerted on the track. Sensors track metabolic rates. The result is two sets of data that can speak to the fairness of competing with a prosthetic limb—"fairness" being understood as "giving no advantage over a natural limb."

I spoke about this process with Alena Grabowski, an expert on biomechanics at the University of Colorado who worked on the cases filed by Pistorius and Rehm with the IAAF. She told me it was very difficult to interpret the studies' results, leaving room for legitimate disagreement. In the case of long jumping, prosthetics limit a long jumper's top speed compared with biological limbs, which is a disadvantage. But the prosthetic limb also appears to facilitate the transfer of vertical force into vertical and horizontal force at takeoff, which is advantageous. Grabowski was convinced Rehm had no advantage; other scientists were not. What all the scientists did agree on, however, was the comparative framework used to assess fairness: artificial versus natural; that is, nonbiological versus biological.

And yet, a regulatory standard of complete natural purity would be incompatible with other essential values of sport, most importantly that of inclusivity. "We need to think more in terms of inclusion instead of exclusion," Grabowski said. "People always believe that the prostheses, because they're different, give an advantage. But they don't. What about contact lenses?" It's hard to argue with her logic. Olympians can use contact lenses and glasses, which are nothing more than prostheses for deficient eyes. The same logic holds for "therapeutic use exemptions," which allow athletes who have illnesses and other conditions to take prohibited substances. Given that sports already allows for artificial aids, excluding artificial limbs simply because they are artificial would create a double standard, turning a reasonable ideal of naturalness into something like a religious purity test, one that runs the risk of being far too exclusive. Better, instead, to compromise on natural purity for the sake of inclusivity. An appropriate approach to natural talent in sports allows athletes who require medication or wear prostheses to compete fairly, while continuing to celebrate nature as it occurs in human physiology.

The value of inclusiveness, like fairness, is written into the IAAF's official constitution. One of the organization's primary goals is "to strive to ensure that no gender, race, religious, political or other kind of unfair discrimination exists, continues to exist, or is allowed to develop in Athletics in any form, and that all may participate in Athletics regardless of their gender, race, religious or political views or any other irrelevant factor."[11]

That gender shouldn't affect one's ability to participate in athletics is now taken for granted, but only after overcoming centuries of pseudoscientific sexism arguing that women were naturally unfit to compete. In ancient Greece, women could not participate in the Olympics, and married women were prohibited from watching. They were also left out of the first modern Olympics, since, in the words of Baron Pierre de Coubertin, founder of the games, their inclusion would be "impractical, uninteresting, unaesthetic, and incorrect."[12]

Even after women were allowed to participate, it was only in those sports believed to accord with their naturally delicate physiology: tennis, croquet, sailing, and golf. Experts warned that more strenuous events might cause women to age prematurely, their uteruses to fall out, and perhaps turn them into men. When the 800 meters was opened to women in the 1928 Olympics, scandalized journalists exaggerated or invented the fatigue experienced by the competitors. "Below us on the cinder path were 11 wretched women, 5 of whom dropped out before the finish, while 5 collapsed after reaching the tape," claimed one account in the *New York Evening Post*. "It is obviously beyond women's powers of endurance, and can only be injurious to them," asserted another writer in the *Montreal Daily Star*.[13] In fact, only nine women had run the race, all of them finished, and only one could conceivably be characterized as collapsing.[14] But the truth didn't matter. In accordance with an invented version of natural law, women were banned from the 800 meters until 1960.

Similar forms of bias remain firmly in place. The women's ski jump did not get added to the Olympics until 2014, and not everyone approved. "If a man gets a serious injury, it's still not fatal, but for women it could end much more seriously," complained Alexander Arefyev, a Russian skiing coach, upon hearing of the decision. "Women have another purpose—to have children, to do housework, to create hearth and home."[15]

Yet, when it comes to women's participation in sports, there's an important distinction to be drawn between two types of discrimination. The first type of discrimination bars women from participating in sports thought to be incompatible with women's biology, such as ski jumping, the 800 meters, and boxing (men's only until the 2012 Olympics). This type of discrimination has been repeatedly shown to have no basis in science. When it comes to women's ability to participate in and excel at any sport, gender should be considered an "irrelevant factor," as the IAAF describes it.

The second type of discrimination is that which divides men and women for the purposes of competition. With the exception of equestrian events and sailing, in every Olympic sport, and in nearly every professional sport, men compete against men and women compete against women. Some have suggested that having men's and women's categories also represents an unfair form of discrimination, and ought to be replaced with different classificatory categories that more accurately reflect the physical traits demanded by a given sport, a practice that already has precedent in the use of weight classes. "For example, for a 100m sprinter, the ideal athlete would perhaps be made up of muscle mass and fast-twitch fibres," writes Roslyn Kerr, a sociologist of sport, "Therefore, rather than classifying by sex, sprinters could be classified by their level of muscle mass and fast-twitch fibres."[16]

Despite such critiques, advocates of female participation in sports generally recognize the need for, and benefits of, sex segregation. The exercise physiologist Ross Tucker puts it straightforwardly: "Being genetically male is the single biggest performance advantage in sport."[17] The advantage enjoyed by biological males exceeds that of other comparatively advantageous traits, including height and weight. A 2010 study quantified the gap between men's and women's top performances in eighty-two different events, from swimming to speed skating. Starting in 1896, the gap narrowed significantly over time as women were allowed to participate in sports. But by 1983 the gap stabilized "at a mean difference of 10.0% ± 2.94 between men and women for all events." The gap depends on the sport, from 5.5 percent for 800-meter freestyle swimming to 36.8 percent for weight lifting. Nevertheless, the overall conclusion is clear: "Results suggest that women will not run, jump, swim or ride as fast as men."[18] Discrimination of the second type is based on good science, not pseudoscientific sexism, and there's a very strong case to be made that it is beneficial for elite female athletes, who would not otherwise get to compete at the highest levels of their sport.

However, policing the division between men and women in sports has a long and fraught history. Since men have the biological advantage, the only athletes subject to sex testing have been women. In the 1960s, when official testing standards were first adopted by the International Olympic Committee and the IAAF, female athletes were subject to incredibly humiliating inspections, including being paraded naked in front of doctors who would inspect their genitalia and pronounce them genuine women. Widespread indignation led to the adoption of chromosome testing, but that proved equally controversial. Unlike weight and height, biological sex occasionally defies simple forms of measurement. This fact was vividly and tragically illustrated in a horrific ordeal endured by the Spanish hurdler Maria José Martínez-Patiño. In 1985, she was looking forward to competing in the World University Games in Japan as a woman, just as she always had. Now a physician, Martínez-Patiño has made public the painful details of what happened. "Our team doctor told me—in front of the team mates I sat with on the night before my race—that there was a problem with my result," she recalls. The doctor told her to fake an injury and withdraw from the race. She agreed, devastated, not knowing what exactly had gone wrong. "Did I have AIDS? Or leukaemia, the disease that had killed my brother?"[19]

Two months later, the official results arrived. She was 46, XY—the male karyotype. But because of a condition known as androgen insensitivity, Martínez-Patiño was insensitive to testosterone, which is why no one, including her, had any idea: "When I was conceived, my tissues never heard the hormonal messages to become male." Eventually her story was leaked to the press, with catastrophic results:

> I was expelled from our athletes' residence, my sports scholarship was revoked, and my running times were erased from my country's athletics records. I felt ashamed and embarrassed. I lost friends, my fiancé, hope, and energy. But I knew that I was a woman, and that my genetic difference gave me no unfair physical advantage. I could hardly pretend to be a man; I have breasts and a vagina. I never cheated.[20]

Martínez-Patiño successfully appealed her disqualification, and after a few different attempts to standardize the testing practices, the IOC gave up and adopted a "suspicion-based" approach. If someone seemed like they might not be a woman, well, then they'd be subjected to further testing.

Unsurprisingly, this approach failed miserably. In 2009, an eighteen-year-old South African runner named Caster Semenya won gold at the Berlin World Championships, crushing her rivals in the once-forbidden-to-women 800 meters. Some of them were suspicious. "These kind[s] of people should not run with us," stated the Italian sixth-place finisher Elisa Cusma. "For me, she's not a woman. She's a man."[21] The IAAF responded by requiring tests, and, as in Martínez-Patiño's case, news of the testing leaked to the press. Some members of the media mocked Semenya's "masculine" appearance and called her a hermaphrodite. She reportedly spent two hours with her legs in stirrups to facilitate examination and photographs of her genitalia, and eventually went into hiding, undergoing treatment for trauma.[22]

In the wake of the Semenya debacle, the IAAF issued a new standard for competing as a female, this time based on testosterone. Again, there were problems. The new standard disqualified all female competitors with hyperandrogenism, a rare condition that causes women to have testosterone levels that are in the typically male range, which, some speculate, is what Semenya has. In 2014, testing revealed that the Indian sprinter Dutee Chand was also above the limit set by the IAAF for female competitors. The Sports Authority of India subsequently ruled that Chand "will still be able to compete in the female category in [the] future if she takes proper medical help and lowers her androgen [testosterone] level to the specified range."[23]

Appalled at the thought of having to artificially lower her naturally produced androgen level with medication, Chand appealed her case to the Court of Arbitration for Sport (CAS), the same administrative body that considered Pistorius's case. The naturalness of her hyperandrogenism, as well as the potential side effects of a medical intervention, was central to her case. She argued that any advantage she enjoyed was a "natural genetic gift," and that in no other case do natural physiological advantages disqualify an athlete. "These interventions are invasive, often irreversible and will harm my health now and into my future," she said in a statement to the CAS. "I am unable to understand why I am asked to fix my body in a certain way simply for participation as a woman. I was born a woman, reared up as a woman, I identify as a woman and I believe I should be allowed to compete with other women, many of whom are either taller than me or come from more privileged backgrounds, things that most certainly give them an edge over me."[24]

Chand's case was taken up by numerous experts, including the Stanford bioethicist Katrina Karkazis. "When a man has unusually high levels of testosterone, the next step is a carbon isotope test," she told me. "If it's deemed to be natural, the case is closed. But for women, if it's natural the case is not closed, and you get ushered into more tests." Although Chand won her appeal, the issue is far from settled. In late 2018, the IAAF issued new testosterone limits that would, once again, disqualify Chand and other hyperandrogenous female athletes. The limits sparked outrage, and Caster

Semenya made a rare public statement denouncing them. "I don't like talking about this new rule," she said. "I just want to run naturally, the way I was born. It is not fair that I am told I must change. It is not fair that people question who I am. I am Mokgadi Caster Semenya. I am a woman and I am fast."[25] (As of this writing, Semenya's fate still hangs in the balance. By the time you read these words, it may have been settled.)

The IAAF's clarifying comments are notably unhelpful, lurching between recognition that sport "seeks to celebrate" a combination of "natural talent and sacrifice and determination" while also maintaining that high testosterone levels are a unique natural biological advantage that should be regulated.[26] Part of the dilemma is that the distinct biological advantage enjoyed by men over women cannot be translated into a rule about testosterone. Hyperandrogenous women are not men. They do not exhibit the same kind of dominance in their respective sports that men would. Nevertheless, the question remains open: If testosterone levels fail to capture that advantage, how can regulatory bodies like the CAS fairly adjudicate the division between men's and women's sports?

Some experts think the best solution is to recognize that science cannot clearly distinguish between male and female in the way that it can between weight classes. Roger Pielke Jr., a political scientist and director of the Sports Governance Center at the University of Colorado, calls science, sex, and gender "the wicked problem," and believes sports organizations should refrain from drawing scientific lines between men and women. Instead, participation would be determined "initially by the athlete in the first instance of participating in organized national or international elite competition segregated into men's and women's categories."[27] Karkazis takes a similar position, favoring a legal definition of sex. "We need to move beyond policing biologically natural bodies," she and her co-authors argue in the *American Journal of Bioethics*. "It is true that some countries may define sex in different ways, but this variability is not necessarily bad."[28]

Pragmatically, relying on national definitions of sex may be the right move for the IOC and the IAAF, but it doesn't resolve the underlying philosophical issues. Moreover, it fails to answer the related question of how to regulate transgender athletes, whose "biologically natural bodies" do not align with their identified gender. Not all transgender people transition in the same way. Some undergo surgery and hormone treatments; others do not. As a consequence, the legal status of a transgender person might not be determined according to the same criteria that should apply to their participation in sports. A legally transgender female who has undergone no biological transitioning will have the same physiological advantages that a biological male does—advantages that are the reason we divide men's and women's sports to begin with. And a transgender male undergoing hormone therapy will be taking drugs that qualify as performance-enhancing under the current rules.

Like the standards of fairness that govern prosthetic limbs, standards of fairness for transgender athletes depend on a natural comparative class. Joanna Harper is a transgender woman, a competitive distance runner, and a medical physicist who studies the performance of transgender athletes. In a 2015 study she looked at eight competitive runners who transitioned from male to female using hormone treatments. To see whether being transgender gave them an advantage, she looked at their ranking when competing against men, and then at their ranking when competing against women, after hormone therapy. The study showed that their relative ranking remained the same: if they were placing around thirty-fifth in races against men before transitioning, they placed around thirty-fifth in races against women of the same age and ability after transitioning. The comparative class in the study is the women against whom the transgender athletes are competing—that is, biological women. Had Harper found that the transgender runners' relative rankings rose, it would have meant that transgender women have an advantage over biological women.[29]

Harper has also studied Tifanny Abreu, a Brazilian transgender volleyball player who will likely compete in the 2020 Olympics. Again, the comparative class when it comes to fairness is biological women. Does Abreu's being transgender give her an advantage over biological women? In some ways it does, Harper believes, but in other ways it does not. "[Hormone therapy] reduces muscle mass, but not to typical female averages," she said in an interview for the *New York Times*. "On average, transgender women are taller, bigger and stronger. For many sports, including volleyball, these are advantages."[30] But transgender women have unique disadvantages as well, since their larger size is compromised by reduced muscle mass and aerobic capacity. Advantages and disadvantages—the same conclusion Alena Grabowski came to about prostheses.

At this point, the claim that no transgender woman could *ever* compete fairly against biological women is clearly born of dogmatism, akin to saying that a prosthetic limb is *necessarily* superior to a biological limb. The claim would need to be backed up with solid scientific evidence that no matter what kind of interventions a transgender woman undergoes, and no matter her age when she transitions, she will *always* enjoy the significant biological advantages that underpin the division between men's and women's sports. There is no such evidence. To the contrary: the evidence we have indicates the interventions that transgender women undergo when transitioning, such as hormone therapy, significantly reduce their biological advantages.

But it would be dogmatism of a different kind to insist on ignoring the very real advantages of biological men over biological women in sports. The right question to ask is what standards should be used to ensure that transgender women are competing fairly—the same question we ask of the standards that govern athletes with prosthetic limbs. More research like Harper's (who is currently consulting with the IOC on regulatory issues regarding transgender athletes) will hopefully provide a better answer. What we know for certain is that biological—natural—sex will be the comparative class in that research.

In the end, the regulatory paradoxes I observed in natural bodybuilding were present in all sports. They are the same paradoxes that arise when designing a national park; when connecting to previous generations of childbearing women; when seeking an unmediated relationship to food; whenever we make laws and rituals out of natural origins. We are truly unnatural animals, simultaneously defined by the artifice of our existence and the significance of our nature. Our home is made in an uncertain borderland, but it is no less sacred for it. And so, when envisioning the future of this home and all its residents, we would do well to remember the lesson of sports: the art of celebrating humanity and nature depends on rejecting the dogmatism of mythic binaries and having the courage to embrace paradox.

Salvation

PARADOX AND UNCERTAINTY ARE FRIGHTENING. Ambiguity about existential questions can feel hopeless and disempowering, a short step from complacency or nihilism. This is why religions tend to emphasize certainty. Divine mysteries and metaphysical riddles exist, but they usually resolve into concrete answers: myths, rituals, and laws that supply stable meaning, clear guidance, and eternal truths—a map of where we came from and where we ought to go.

When it comes to the precarious future of our relationship with nature and naturalness, the need for stable meaning and clear guidance is especially acute. Every question is existential. As I write these words, lab-grown meat is on the cusp of becoming a fixture in restaurants and supermarkets—at least if consumers are willing to accept its unnatural origins in giant steel vats. The gene-editing technology called CRISPR-Cas9 has already been used to splice woolly mammoth gene sequences into an elephant's DNA, a first step toward the eventual goal of "de-extinction." With our choices and our votes, humankind is poised to make monumental decisions about the meaning and importance of natural goodness. There is a distinct possibility that thirty years from now, I will be ordering a sustainably lab-grown mammoth steak as I discuss with my daughter whether she should purchase disease-resistant genes for her future baby or opt for a natural birth instead. Climate change is increasing the occurrence of infectious diseases, she tells me, and it's bound to get worse. Maybe the genes are worth it?

A few years ago, when I began my research, I would have dismissed the value of naturalness entirely. (Eat the lab-grown mammoth steaks! Go for

the modified baby!) At the time, my skepticism about faith in nature's good-
ness had become its own kind of faith, a photonegative of the false ideology
I sought to discredit. Natural is not perfect; therefore, it is meaningless.
Favoring a choice because it is natural amounts to a superstitious mistake.
My faith had its own derisive mantras, shared frequently on social media:
"Watch out for dihydrogen monoxide in your water!"; "Cyanide is natural!";
and so on. Scientific truths were cited with the decisive zeal of a fire-and-
brimstone pastor citing scripture.

At first, the righteous certainty was empowering. I set out to confirm my
biases and found, happily, that most studies fail to establish clear health benefits for eating organic, and what benefits there might be are negligible in
comparison to other lifestyle choices. I learned that organic farming can and
does use pesticides, contrary to popular belief. (The hypocrisy of puritanical
faiths is exhilarating to a nonbeliever.) I learned that some organic vegetables owe their existence not to breeding, but rather to seeds that have been
bombarded with mutagenic radiation, a far cry from the natural genealogy
invoked by their non-GMO butterfly label. I raged against fearmongering
Facebook posts that blamed cancer and autism on conventional agriculture
and GMOs.

Eventually I set up an interview with an expert on pesticides, certain his
shopping habits would further confirm my findings.

"I bet you don't waste your money on organic, right?" I asked. "No
health difference!"

"Actually, I buy organic bananas," he said. "The pesticides used in non-
organic can harm the workers. Not about my health, though. Depends on
what I know about how the crop is grown."

I was stunned. Without meaning to, this chemist had laid bare two terrible problems in my thinking. The first was that I hadn't been considering
the welfare of farmworkers. Or, rather, I had only been considering myself.
All of my initial research, the deep dives into data—I never gave a second
thought to the people laboring to produce our food; nor, for that matter,
did I consider the possible benefits of organic farming to soil quality or biodiversity. I knew these were important questions, but I managed to ignore
them. My primary concern was safeguarding my health—and my ideology.

The second problem was the myopic dogmatism I'd been trying to protect. Confronted with a false faith, I had resolved that it was wholly evil. If the
nature-worshipers reflexively opposed anything deemed unnatural—GMOs,

lab-grown meat, synthetic vanilla flavor—well, I supported it. This reactionary attitude, I later learned, has an analogue in the early days of the organic food movement. Organic advocates made extravagant claims about the "the methods of Nature—the supreme farmer," and the existential risk of violating her eternal laws.[1] "Artificial manures lead inevitably to artificial nutrition, artificial food, artificial animals, and finally to artificial humans," warned the botanist Sir Albert Howard in 1940, his book tellingly titled *An Agricultural Testament*.[2] As science journalist Charles C. Mann describes it, the movement saw "industrial agriculture as a threat to both the social and divine order."[3]

The scientists behind industrial agriculture pushed back against the "muck and magic" of their critics, accusing them of cultish pseudoscience. The result, which Mann chronicles in his book *The Wizard and the Prophet*, was nothing short of a religious war—"chemicalist versus organiculturist"— that precluded any possibility of productive dialogue. Those with faith in the natural order, the prophets, saw the technological wizards as godless heathens. And the wizards sought only to discredit the prophets—just as I did when I began my research.

Ironically, rigid theological binaries result in something that advocates of organic agriculture have long warned against: monoculture. In the case of crops, monoculture means planting a single crop in the same field, year after year. Though productivity is often improved, the practice frequently leads to depleted soil nutrients and higher vulnerability to pests, disease, and weather-related disasters. Diversity of crops, on the other hand, can foster a more flexible, resilient farming operation, with healthier soil, more biodiversity, and better prospects for long-term survival.

In the case of thinking about nature, the problem is *ideological monoculture*. Simplicity and homogeneity take precedence over diversity, complexity, and change. Righteous laws and rituals are universal. Disobedience is sacrilege. Disagreement is heresy. Compromise is hypocrisy. Technology will damn us or save us.

The alternative to monoculture is polyculture, the deliberate cultivation of diversity for the sake of long-term resilience. But here is where the analogy between agriculture and ideology breaks down. Polyculture is more like nature, say its supporters, so polyculture is always good and monoculture is always bad. By contrast, *ideological polyculture* refuses eternal certitudes and mythic binaries based on naturalness. Never say never; never say always. No foundational principles about Nature's laws predetermine your position on genetically modifying babies or proper sexual habits. The default is uncertainty, ambiguity, and openness to complexity and change.

It may seem as if this default position can only lead to paralysis and indecision. In the face of urgent threats to the natural world and questions about the very definition of humanity, such indecision would be gravely immoral. But throughout my years of research for this book, I discovered that passionate activism is completely compatible with acknowledging complexity and ambiguity. From Doug Smith in Yellowstone to Joanna Harper on transgender athletes, I met proactive people for whom knowledge and experience were directly correlated with fewer overarching certainties about the meaning of nature and how best to honor it. Their uncertainty is a virtue, not a shortcoming, when it comes to improving the world.

There is no better example of how to live out this virtue than the environmental historian and activist Nancy Langston. Langston has spent much of her distinguished career documenting the dangers of toxins and fighting to keep them out of our bodies and our ecosystems. Her 2010 book *Toxic Bodies* is a grim indictment of the government's failure to regulate endocrine disruptors, and her latest, *Sustaining Lake Superior*, chronicles the destructive flow of industrial pollutants into a lake she has loved and defended for over seventeen years. You'd be hard pressed to find someone more familiar with the harmful effects of synthetic chemicals, or more devoted to protecting humanity and nature from future catastrophes.

My first conversation with Langston covered a range of topics, all of which circled around concern for the nonhuman world and our relationship with it. As we talked about the urgency of fighting climate change, she suddenly laughed.

"I'm saying all this while I get ready to put on protective gear to deal with glossy buckthorn and spotted knapweed! Not exactly natural, is it?"

Her property at the time included a large area that was being overgrown by invasive plant species. The gear was necessary, she explained, because of the highly toxic herbicides needed to eradicate them. After our phone call she would become a warrior in a hazmat suit, dousing plants with poison— with the end goal of promoting biodiversity and returning native species to the ecosystem.

It's a shocking image if you associate goodness with purity. But Langston is no dogmatist. Decades of work have resulted in a humble wisdom that abhors taboos and binaries. Repeatedly she cautioned against simple, static answers to big questions. During our conversations she ticked off a list of her nuanced beliefs, many of which have changed over time as she acquired more evidence: *GMOs have benefits (synthetic insulin for diabetics) and drawbacks (unforeseen ecological effects when rashly introduced). Nuclear energy should be phased out eventually, but only when fossil fuels won't replace it. The value of synthetic chemicals depends on the chemical and the context.* While discussing these examples she was remarkably circumspect, and stressed the need to know more before coming to a definitive conclusion in any given circumstance.

As if to illustrate her point, between our conversations Langston moved to a new house closer to Lake Superior. There she continued her fight against spotted knapweed, but traded Roundup for manual weeding because she judged it too risky to use the herbicide so close to the water. Her decision was not made according to general principles, but rather local circumstances. In the same spirit, she reminded me that it's probably unfair to paint all invasive plant species with the same brush. They, too, need to be considered on a case-by-case basis—the mere fact of their nonnative status does not make them evil. "I remember walking around Lake Superior, this incredible pristine lake," she told me. "And I was looking at all the invasive species and thought, Who cares if the lupine isn't native! It's an extraordinary sign of the northwest spring and I plant it everywhere."

There's a rhythm to her thinking that comes through when we talk, from instability to stability and back again. Stability results in action. But the default position is instability, the dynamism of change instead of philosophical rigor mortis.

Langston is relentlessly concerned with the welfare of human and nonhuman creatures. She has spent years of her life devoted to them, hearing their concerns, learning how they function, coming to understand what threatens them and what allows them to flourish, trying to grasp the complex systems that sustain them. As a consequence, she ties her criteria for goodness to specific contexts: *this* lake, *these* fish, *this* community of people. The only eternal answer is to begin by asking questions, of humans and nonhumans, of individuals and systems. It's remarkable humility from a person fully qualified for certainty. Does she value nature, I asked? Is naturalness a good thing? "I haven't given up on reverence for nature," she answered, haltingly. "I'm just philosophically confused."

Philosophically confused. It sounded like a confession when she said it, an admission of guilt. I was reminded of a similar confession, this one from Joel Salatin of Polyface Farm. After our communal dinner of pastured pork he announced a new "enterprise" to the workers and guests. In response to high demand for soy-free chickens (the chickens get most of their calories from locally sourced non-GMO soy), they were going to start raising some chickens on a special feed made from fishmeal, flaxseed oil, and Austrian peas. When I asked how feeding fish to chickens fit with "nature's template," Salatin gave me a resigned look. "I'm a hypocrite, it's true," he said. "We all choose our compromises." Then he smiled ruefully. "At least I'm honest about mine."

But compromising nature's template doesn't need to be synonymous with hypocrisy. Philosophical confusion about nature isn't something to feel guilty about. These are only sins if we believe the ideal position regarding nature is faith in its goodness, and certainty about what that goodness should look like. That kind of faith leads to rigidity and restrictiveness—to ideological monoculture, an inability to dialogue, and simplistic top-down approaches to issues that require nuanced, local, bottom-up answers.

By contrast, if paradox and uncertainty are the default position—if the *truth* about our relationship to nature is paradoxical and uncertain—then compromise is not hypocrisy, but rather a necessity. Philosophical confusion isn't a sin; it's a virtue. I don't know whether I would want my daughter to genetically modify her children. How could I, without knowing what she is like as an adult? Without knowing what society is like in the future? What if I am thinking about it the wrong way? What if the problem with modifying humans isn't the artifice of it, but unequal access to it? Only prophets and wizards could have the wisdom and the confidence to pronounce on these issues so far in advance. I am neither.

Dogmatic religiosity does not encourage uncertainty about existential questions. Laws are eternal. Rituals purify. Myths tell of divinity creating order from chaos. The story of creation in the Bible describes how plants and animals came into being, as well as the proper relationship of humans to them: "They will have power over the fish, the birds, and all animals, domestic and wild, large and small." The story of sacrifice in the ancient Purusha Sukta, Hymn 10.90 from the Rigveda, provides certainty about the divine origin of the ideal social order.

But certainty is not the only choice. Now, at the end of a journey, I find myself at the beginning of a new one. I am more philosophically confused about nature than I was when I began. Maybe you feel the same way, full of questions instead of answers. This is no reason for shame or guilt. It is not something to be overcome. Uncertainty is humility, and humility can also be sacred, its own source of rituals and laws, which, like nature, can change while remaining true to themselves.

How to accept this way of being instead of fearing it? Remember: the end is hidden in the beginning. Origin stories create identity. There are many stories. What happens to naturalness if we choose a different one? What if we, the unnatural animals, begin with a hymn to uncertainty and paradox? What will we become?

THE NASADIYA SUKTA, HYMN 10.129 OF THE RIGVEDA

Then even nothingness was not, nor existence,
There was no air then, nor the heavens beyond it.
What covered it? Where was it? In whose keeping?
Was there then cosmic water, in depths unfathomed?

Then there was neither death nor immortality
nor was there then the torch of night and day.
The one breathed windlessly and self-sustaining.
There was that one then, and there was no other.

At first there was only darkness wrapped in darkness.
All this was only unillumined cosmic water.
That one which came to be, enclosed in nothing,
arose at last, born of the power of heat.

In the beginning desire descended on it—
that was the primal seed, born of the mind.
The sages who have searched their hearts with wisdom
know that which is kin to that which is not.

Their cord was stretched across.
Was there below? Was there above?

There were seminal powers; there was fertility.
Below was impulse, above was giving-forth.

But, after all, who knows, and who can say
whence it all came, and how creation happened?
The gods themselves are later than creation,
so who knows truly whence it has arisen?

Whence all creation had its origin,
the creator, whether he fashioned it or whether he did not,
the creator, who surveys it all from highest heaven,
only he knows—or if he does not know?[4]

Acknowledgments

I COULD NEVER HAVE WRITTEN THIS BOOK without the help and guidance of countless individuals. To everyone who was kind enough to provide information and interviews, often at length and sometimes repeatedly, your expertise and your patience with my lack of it, has been invaluable: Rebecca Altman, Chelsey Armstrong, Susie Bautista, Nadia Berenstein, Jennifer Bernstein, Nick Blurton-Jones, Adam Boyette, Matthew Brignall, Catherine Cameron, Ken Cameron, John Coupland, Michael Cournoyea, Sidney Dekker, Colleen Derkatch, Michael Egan, Robert Gastaldo, Robert George, Brenna Hassett, Keira Havens, Kristen Hawkes, Barry Hewlett, Roald Hoffmann, Brooke Holmes, Kingsley Ighobor, Carolina Izquierdo, Jason Kawall, Helen King, Mel Konner, Nancy Langston, Renee Pennington, Susan Pfeiffer, Roger Pielke, Chelsea Polis, Jennifer Reich, Michael Reid, David Ropeik, Karen Rosenberg, Paul Rozin, Richard Sachleben, Palmira Saladie, Paul Schullery, Anna Shoemaker, Ken Tankersley, Wenda Trevathan, Amy Tuteur, Barbara Ulrich, Paul Unschuld, Connie Walker, Cathy Whitlock, and Patti Zettler.

During my travels, Filip van Noort, Judith Westerink, and Annette Haberink helped me navigate the Netherlands; Scott guided us with wisdom in Peru; Doug Smith, Ilona Popper, Ashea Mills, and Michael Tercek made Yellowstone sacred; Sayaka Nakai made me feel at home in Tokyo, and Makiko Segawa showed me Fukushima (a story still to be written!); Todd Allen helped me learn about nuclear energy (same untold story); Brooke Crowley introduced me to the bonobos, Vern Scarborough and Ken Tankersley shared their insights. Joel Salatin was gracious. Robert Sapolsky and Charles Mann explained how to write with humility (if I failed it wasn't their fault!). The support of James Madison University has been crucial:

what a lucky place to have landed. So has the support of the Breakthrough Institute, without which I never would have met so many of the people who informed my research and shaped my thinking—Alex Trembath and Ted Nordhaus, thank you for inviting me along. To all my editors who believed a religious studies professor could write about science, I'm in your debt, especially Laura Helmuth, who believed it before there was much evidence. Jen Gunter and Tim Caulfield: you took me under your wings early in the game, and I'm forever grateful. Sam Haselby and Maria Godoy let me write for them about these ideas when they were still rough. Rachel Laudan, Jim Hamblin, and Mark Schatzker have been wonderful friends and conversation partners. Peyton Upshaw went above and beyond. Alex Hsu found the hymn. Daniel Morgan translated instantly. Rick Rosengarten taught me how to think about religion and stories. My editors Amy Caldwell and Ed Lake believed in the book. Brian Baughan and Susan Lumenello polished it well. My agent, Anna Sproul-Latimer, has been endlessly supportive and an amazing human being, every author should be so lucky.

Hazel: sorry kiddo, I know, your dad is tough when he's writing. My parents: as they say, without them I wouldn't exist—and neither would this book, which they have made possible in so many ways. And Paris: I don't have the words, but I think you know.

Notes

INTRODUCTION

1. Timothy Egan, "Fake Meat Will Save Us," *New York Times*, June 21, 2019, https://www.nytimes.com/2019/06/21/opinion/fake-meat-climate-change.html.
2. Pollan, *The Omnivore's Dilemma*, 126.
3. Salatin, *The Marvelous Pigness of Pigs*, 48.
4. Pollan, *The Omnivore's Dilemma*, 245.
5. Darwin, *The Works of Charles Darwin*, vol. 10, *The Foundations of the Origin of Species*, 40.
6. Darwin, *On the Origin of Species*, 84, 489. For nature capitalized and personified, see, e.g., 480.
7. Comments can be accessed at "Use of the Term Natural on Food Labeling," https://www.fda.gov/food/food-labeling-nutrition/use-term-natural-food-labeling.
8. Pride Chigwedere et al., "Estimating the Lost Benefits of Antiretroviral Drug Use in South Africa," *JAIDS (Journal of Acquired Immune Deficiency Syndromes)* 49, no. 4 (2008): 410–15.
9. Michael Specter, "The Denialists," *New Yorker*, March 12, 2007, https://www.newyorker.com/magazine/2007/03/12/the-denialists.
10. Reich, *Calling the Shots*, 98.
11. Data on energy access from the International Energy Agency's analysis can be found here: https://www.iea.org/sdg.
12. Thoreau's locomotive was the image chosen for the cover of Leo Marx's classic study of nature invaded by technology, *The Machine in the Garden*.
13. Mill, *Three Essays on Religion*, 3. Note Mill's essay on nature was in a volume dedicated to the study of religion.
14. The "appeal to nature fallacy" is often used interchangeably with the "naturalistic fallacy," a different philosophical fallacy coined by the philosopher G. E. Moore, which refers to the impossibility of defining goodness in terms of natural properties. See, for instance, Steven Pinker's discussion in *The Blank Slate*, 150.

PART I: MYTH

1. Portions of this introduction appeared in Alan Levinovitz, "The Curse of Frankenstein," *The Conversation*, May 20, 2015, https://theconversation.com/the-curse-of-frankenstein-how-archetypal-myths-shape-the-way-people-think-about-science-42077.
2. Kia Shant'e Breaux, "Five Piglets Join Dolly as Clone Farm Grows," Associated Press, March 14, 2000, https://www.greensboro.com/five-piglets-join-dolly-as

-clone-farm-grows-researchers-hope/article_373e365f-9c78-5e9b-8d23-4b2dd
03c87e8.html.

3. Paul Lewis, "Since Mary Shelley," *New York Times,* June 2, 1992, https://www.nytimes
.com/1992/06/16/opinion/l-mutant-foods-create-risks-we-can-t-yet-guess-since
-mary-shelley-332792.html.

4. *Times* Wire Service, "Dr. Patrick Steptoe; Gynecologist Delivered the First Test-
Tube Baby," *Los Angeles Times,* March 23, 1988, https://www.latimes.com/archives
/la-xpm-1988-03-23-mn-1905-story.html.

5. Lakoff and Johnson, *Metaphors We Live By,* 4.

6. Cited in Lincoln, *Theorizing Myth,* 41–42.

CHAPTER 1: IN THE BEGINNING

1. See the "About" page on Lamaze International's website, https://elearn.lamaze.org
/pages/about-our-classes.

2. See the "Myths about Lamaze" page, https://www.lamaze.org/myths-about lamaze.

3. Bradley, *Husband-Coached Childbirth,* 102.

4. Bradley, *Husband-Coached Childbirth,* 239.

5. Bradley, *Husband-Coached Childbirth,* 239.

6. Genevieve Howland, *The Mama Natural Week-by-Week Guide to Pregnancy and Childbirth* (New York: Gallery Books, 2017), 392; Ina May Gaskin, *Ina May's Guide to
Childbirth* (New York: Bantam, 2017), 270; Judith Lothian and Charlotte DeVries,
Giving Birth with Confidence: Official Lamaze Guide, 3rd ed. (Boston: Da Capo Life-
long Books, 2017), chap. 5, "Nourishing Your Body."

7. Heidi Murkoff and Sharon Mazel, *What to Expect When You're Expecting,* 5th ed.
(New York: Workman Publishing Co., 2016), 84, 88, 89, 132, 192.

8. Lothian and DeVries, *Giving Birth with Confidence,* chap. 7, "The Simple Story of
Birth."

9. Allie Einstein and Hannah Ferrett, "Mom of 3 Gets Naked, Squats in Stream, Gives
Birth to Next Baby," *Sun,* July 13, 2016. Accessed in *New York Post:* https://nypost
.com/2016/07/13/mom-of-3-gets-naked-squats-in-stream-gives-birth-to-next
-baby.

10. Gaskin, *Ina May's Guide,* 99.

11. Ramiel Nagel, *Healing Our Children: Sacred Wisdom for Preconception, Pregnancy, Birth
and Parenting* (Los Gatos, CA: Golden Child Publishing), 51.

12. Rush, *Medical Inquiries and Observations,* 61.

13. Leclerc, *A Natural History, General and Particular,* 272, in a section entitled "Varieties
of the Human Species."

14. Thomas Huxley, *T. H. Huxley on Education: A Selection from His Writings* (Cambridge,
UK: Cambridge University Press, 1971), 68–69.

15. On this point, see Bush, *Slave Women in Caribbean Society,* 14–15.

16. Kelly M. Hoffman et al., "Racial Bias in Pain Assessment and Treatment Recommendations, and False Beliefs about Biological Differences between Blacks and
Whites," *Proceedings of the National Academy of Sciences* 113, no. 16 (2016): 4296–301.

17. Candace Johnson, "The Political 'Nature' of Pregnancy and Childbirth," *Canadian
Journal of Political Science/Revue canadienne de science politique* 41, no. 4 (2008):
889–913, 908.

18. Bradley, *Husband-Coached Childbirth,* 9.

19. Gaskin, *Ina May's Guide*, 141–42.
20. CIA World Factbook, "Maternal Mortality Rate," https://www.cia.gov/library /publications/the-world-factbook/rankorder/2223rank.html.
21. Johnson, *All Natural**, 27.
22. Iliana V. Kohler et al., "Comparative Mortality Levels among Selected Species of Captive Animals," *Demographic Research* 15 (2006): 413–34. Cited by Amy Tuteur, "Mothering Like an Animal," *Skeptical OB*, November 20, 2017, https://www .skepticalob.com/2017/11/mothering-like-an-animal.html. Emphasis added.
23. Leanne T. Nash, "Parturition in a Feral Baboon (*Papio anubis*)," *Primates* 15, nos. 2–3 (1974): 279–85.
24. Nga Nguyen et al., "Comparative Primate Obstetrics: Observations of 15 Diurnal Births in Wild Gelada Monkeys (*Theropithecus gelada*) and Their Implications for Understanding Human and Nonhuman Primate Birth Evolution," *American Journal of Physical Anthropology* 163, no. 1 (2017): 14–29.
25. Thomas N. Headland and Janet D. Headland, "Limitation of Human Rights, Land Exclusion, and Tribal Extinction: The Agta Negritos of the Philippines," *Human Organization* 56, no. 1 (1997): 79–90.
26. Susan Pfeiffer et al., "Discernment of Mortality Risk Associated with Childbirth in Archaeologically Derived Forager Skeletons," *International Journal of Paleopathology* 7 (2014): 15–24.
27. Hewlett and Lamb, *Hunter-Gatherer Childhoods*, 97.
28. Bernardo Arriaza et al., "Maternal Mortality in Pre-Columbian Indians of Arica, Chile," *American Journal of Physical Anthropology* 77, no. 1 (1988): 35–41.
29. Charlotte L. King et al., "Let's Talk about Stress, Baby! Infant-Feeding Practices and Stress in the Ancient Atacama Desert, Northern Chile," *American Journal of Physical Anthropology* 166, no. 1 (2018): 139–55.
30. Angela R. Lieverse et al., "Death by Twins: A Remarkable Case of Dystocic Child-birth in Early Neolithic Siberia," *Antiquity* 89, no. 343 (2015): 23–38.
31. Dillard, *Pilgrim at Tinker Creek*, 176.
32. Lothian and DeVries, *Giving Birth with Confidence*, chap. 1, "Having a Safe, Healthy Birth."
33. Biss, *On Immunity*, 37.
34. Not everyone is convinced that the EEA is a useful or even coherent concept. For a critical discussion of its utility in medicine, see Michael Cournoyea, "Ancestral As-sumptions and the Clinical Uncertainty of Evolutionary Medicine," *Perspectives in Biology and Medicine* 56, no. 1 (2013): 36–52.
35. Hrdy, *Mother Nature*, 535.
36. Hrdy, *Mother Nature*, 535.
37. Janesh K. Gupta et al., "Position in the Second Stage of Labour for Women without Epidural Anaesthesia," *Cochrane Database of Systematic Reviews* 5 (2017).
38. Van Gennep, *The Rites of Passage*, 51.
39. I borrow "story-shaped world" from Wicker, *The Story-Shaped World*.

CHAPTER 2: THE TRUE VINE
1. Frode Alfnes et al., "Consumers' Willingness to Pay for the Color of Salmon: A Choice Experiment with Real Economic Incentives," *American Journal of Agricul-tural Economics* 88, no. 4 (2006): 1050–61.

2. Translated with small changes from the Mexican government's version, found here: https://www.gob.mx/cms/uploads/attachment/file/104879/DO_Orgullo_de _Mexico.pdf.

3. Eric Schroeder, "McDonald's Makes Changes to Vanilla Soft Serve," *Food Business News,* May 18, 2017, https://www.foodbusinessnews.net/articles/9362-mcdonald-s -makes-change-to-vanilla-soft-serve.

4. Pollan, *In Defense of Food,* 148. It's important to note that Pollan's treatment here is conscious of the tensions built into making this kind of rule. His first book, *Second Nature,* is one of the best on the contradictions of valuing naturalness that I have ever read.

5. The following section synthesizes information from various sources: Cameron, *Vanilla Orchids*; Havkin-Frenkel and Belanger, *Handbook of Vanilla Science and Technology*; Odoux and Grisoni, *Vanilla,* especially chap. 16, Patricia Rain and Pesach Lubinsky, "Vanilla Use in Colonial Mexico and Traditional Totonac Vanilla Farming," and chap. 17, Raoul Lucas, "Vanilla's Debt to Reunion Island"; Rain, *Vanilla*; Ecott, *Vanilla*; Pesach Lubinsky, "Historical and Evolutionary Origins of Cultivated Vanilla," PhD diss., University of California Riverside, 2007.

6. Cited in Woloson, *Refined Tastes,* 55.

7. Woloson, *Refined Tastes,* 55.

8. United States Department of Agriculture, "Notice of Judgment 5: Misbranding of Vanilla Extract," *Notices of Judgment under the Food and Drugs Act, Issues 1–250* (Washington, DC: USDA Office of the Secretary, 1908). See also "Notice of Judgment 301," for misbranding of vanilla flavor, where the plaintiff, St. Louis Coffee & Spice Mills, prevailed by exploiting a technical distinction between "extract" and "flavor."

9. Alvin W. Chase, *Dr. Chase's Recipes or Information for Everybody,* 2nd Canadian ed. (London, Ontario: E. A. Taylor, 1871), 54–56. My personal copy appears no different from earlier editions found online, which date back to at least as far as 1864.

10. Information in this section from Pendergrast, *For God, Country and Coca-Cola.*

11. Pendergrast, *For God, Country and Coca-Cola,* 23.

12. Pendergrast, *For God, Country and Coca-Cola,* 65.

13. The Stepan Company, which runs the New Jersey factory, is the only lawful importer of coca leaves in the US, buying what would amount to millions of dollars of cocaine per year from Peruvian and Bolivian sources.

14. Cited in Wexler, *History of Toxicology and Environmental Health,* 29.

15. Apicius, *Cooking and Dining in Imperial Rome,* trans. Joseph Dommers Vehling, Book IV, Section VI, accessed through Project Gutenberg, https://www.gutenberg.org /files/29728/29728-h/29728-h.htm.

16. Wilson, *Swindled,* 259–60.

17. Nadia Berenstein, "Making a Global Sensation: Vanilla Flavor, Synthetic Chemistry, and the Meanings of Purity," *History of Science* 54, no. 4 (2016): 399–424.

18. Zaria Gorvett, "The Delicious Flavor with a Toxic Secret," *BBC Future,* June 20, 2017, http://www.bbc.com/future/story/20170620-the-delicious-flavour-with-a-toxic-secret.

19. Berenstein, "Making a Global Sensation," 415.

20. "Vanilla Extract," *Cooks Illustrated,* March 2009, archived article accessed at https:// web.archive.org/web/20181012161531/https://www.cooksillustrated.com/taste_tests /455-vanilla-extract.

21. J. Kenji López-Alt, "Taste Test: Is Better Vanilla Extract Worth the Price?," *Serious Eats*, December 16, 2013, https://sweets.seriouseats.com/2013/12/taste-test-is-better-vanilla-extract-worth-the-price.html.
22. Roald Hoffmann, "Fraudulent Molecules," *New Yorker*, July–August 1997.
23. Martin B. Hocking, "Vanillin: Synthetic Flavoring from Spent Sulfite Liquor," *Journal of Chemical Education* 74, no. 9 (September 1997): 1055–59.
24. Danwatch, *The Hidden Cost of Vanilla: Child Labour and Debt Spirals*, December 8, 2016, https://old.danwatch.dk/en/undersogelse/thehiddencostofvanilla.
25. Jonathan Watts, "Madagascar's Vanilla Wars: Prized Spice Drives Death and Deforestation," *Guardian*, March 31, 2018, https://www.theguardian.com/environment/2018/mar/31/madagascars-vanilla-wars-prized-spice-drives-death-and-deforestation.
26. "Eve Explains Evolva's Fermentation," posted on YouTube August 28, 2014, https://www.youtube.com/watch?v=y96w21HkaHQ.
27. Summary of the debate, quotes, and a link to the letter found on Friends of the Earth website: https://foe.org/news/2014-08-haagen-dazs-says-no-to-synbio.
28. Food Babe, "Do You Eat Beaver Butt?," posted on YouTube, September 9, 2013, https://www.youtube.com/watch?v=nweK6VRM8a8.
29. See, e.g., the deep dive by Nadia Berenstein, "A History of Flavoring Food with Beaver Butt Juice," *Vice*, December 20, 2018, https://www.vice.com/en_us/article/a3m885/a-history-of-flavoring-food-with-beaver-butt-juice.
30. David Morgan, "Dr. Mark Hyman Answers the Question: 'Food: What the Heck Should I Eat?,'" *CBS News*, March 19, 2018, https://www.cbsnews.com/news/dr-mark-hyman-food-what-the-heck-should-i-eat.
31. Jean-Jacques Rousseau, *Emile, or Education*, trans. Barbara Foxley, MA (London and Toronto: J.M. Dent and Sons, 1921; New York: E. P. Dutton, 1921), accessed online: https://oll.libertyfund.org/titles/2256#Rousseau_1499_536.
32. Graham, *Treatise on Bread, and Bread-Making*, 14–18.
33. Hoffmann, *The Same and Not the Same*, 118.
34. Hoffmann, *The Same and Not the Same*, 123.
35. John Steinbeck, *Travels with Charley: In Search of America* (1962; New York: Penguin, 1997), 83.

CHAPTER 3: STATES OF NATURE

1. Scott, *Against the Grain*, 5.
2. Scott, *Against the Grain*, 88.
3. Scott, *Against the Grain*, 103.
4. Scott, *Against the Grain*, 107.
5. Pinker, *Enlightenment Now*, 134.
6. Cited in Pinker, *Enlightenment Now*, 37.
7. Krech, *The Ecological Indian*.
8. Kimberly Tallbear, "Shepard Krech's *The Ecological Indian*: One Indian's Perspective," *International Institute for Indigenous Resource Management* (2000): 1–6.
9. "Peaceful Societies: Alternative to Violence and War," https://cas.uab.edu/peacefulsocieties.
10. Keeley, *War before Civilization*, 22.
11. Cameron, *Captives*.
12. Sapolsky, *Behave*, chapters 9 and 17.

13. Michaeleen Doucleff, "Are Hunter-Gatherers the Happiest Humans to Inhabit the Earth?," NPR, October 1, 2017, https://www.npr.org/sections/goatsandsoda/2017/10/01/551018759/are-hunter-gatherers-the-happiest-humans-to-inhabit-earth.

14. Marshall Sahlins, "Notes on the Original Affluent Society," in *Man the Hunter*, ed. R. B. Lee and I. DeVore (Chicago: Aldine, 1968), 85–89.

15. Some parts of this paragraph are slightly modified from Alan Levinovitz, "It Never Was Golden," *Aeon*, August 17, 2017, https://aeon.co/essays/nostalgia-exerts-a-strong-allure-and-extracts-a-steep-price.

16. Rachel Laudan, "Was the Agricultural Revolution a Terrible Mistake? Not if You Take Food Processing into Account," *A Historian's Take on Food and Food Politics*, January 21, 2016, http://www.rachellaudan.com/2016/01/was-the-agricultural-revolution-a-terrible-mistake.html.

17. Timothy N. Bond and Kevin Lang, "The Sad Truth about Happiness Scales," *Journal of Political Economy* 127, no. 4 (2019): 1629–40.

18. Max Roser and Esteban Ortiz-Ospina, "Happiness and Life Satisfaction," *Our World in Data*, May 2017, https://ourworldindata.org/happiness-and-life-satisfaction.

19. Doucleff, "Are Hunter-Gatherers the Happiest Humans?"

20. Patricia Draper, "Remembering the Past: !Kung Life History Narratives," in *Balancing Acts: Women and the Process of Social Change*, ed. Patricia Lyons Johnson (Boulder, CO: Westview Press, 1992), 14.

21. Draper, "Remembering the Past," 13.

22. In his discussion of the history of these debates, Robert Sapolsky suggests that the current state of affairs is a marked improvement over the 1960s and '70s, when anthropology departments were splitting apart and the biologist E. O. Wilson was physically attacked. My own take is somewhat more pessimistic. See Sapolsky, *Behave*, 383–85.

23. Personal correspondence, kept anonymous, October 29, 2017.

24. Though human activity did not pose an existential threat to the environment until the twenty-first century, it has long been noted that human activity can negatively affect the natural world. In the first chapter of the *Mencius*, a Confucian text from c. 300 BCE, the author cautions against overfishing and deforestation; in 1661, John Evelyn published *Fumifugium*, a pamphlet addressing air pollution in London.

25. *The Complete Works of Zhuangzi*, trans. Burton Watson (New York: Columbia University Press, 2013), 123. I have modified the end slightly to capture the use of *ziran* more clearly.

26. Translation and citation from Robert Ford Campany, "The Meanings of Cuisines of Transcendence in Late Classical and Early Medieval China," *T'oung Pao* 91 (2005): 13–14.

27. Claude Levi-Strauss, *The Raw and the Cooked*, trans. John and Doreen Weightman (Chicago: University of Chicago Press, 1983), 59.

28. Eliade, *The Myth of the Eternal Return*, 98.

29. Ellingson, *The Myth of the Noble Savage*, 127.

30. Cited in Ellingson, *The Myth of the Noble Savage*, 51.

31. *Rudyard Kipling's Verse, Inclusive Edition, 1885–1918* (1922; Garden City, NY: Doubleday), accessed online: www.bartleby.com/364/169.html.

32. Cited in Ellingson, *The Myth of the Noble Savage*, 50.

33. *American History Told by Contemporaries*, vol. 1, *Era of Colonization 1492–1689*, ed. Albert Bushnell Hart (1919; Honolulu: University Press of the Pacific, 2002), 38.

34. Scott, *Against the Grain*, 155.
35. Scott, *Against the Grain*, 114.
36. John Lanchester, "The Case against Civilization: Did Our Hunter-Gatherer Ancestors Have It Better?," *New Yorker*, September 18, 2017.
37. Jared Diamond, "Best Practices for Raising Kids? Look to Hunter-Gatherers," *Newsweek*, December 17, 2012, https://www.newsweek.com/best-practices-raising-kids-look-hunter-gatherers-63611.
38. Staff, "How Alicia Silverstone Started Potty Training Her 6-Month-Old Son," *People*, April 19, 2014, https://people.com/parents/how-alicia-silverstone-started-potty-training-her-6-month-old-son.
39. Elizabeth Kolbert, "Spoiled Rotten," *New Yorker*, June 25, 2012, https://www.newyorker.com/magazine/2012/07/02/spoiled-rotten.
40. Glenn H. Shepard, "Gift of the Spider: Spinning, Weaving and Womanhood among the Matsigenka of Peru," https://chacruna.net/gift-spider-woman-spinning-weaving-womanhood-among-matsigenka-peru. First published as "Gift of the Spider Woman: Spinning and Weaving among the Matsigenka," in *Enigmatic Textile Art of the Peruvian Amazon: Ashaninka, Matsiguenka, Yanesha, Yine*, ed. Jaime Valentin Coquis et al (Lima: Cotton Knit S.A.C., 2006), 39–56.
41. Brandon H. Hidaka, "Depression as a Disease of Modernity: Explanations for Increasing Prevalence," *Journal of Affective Disorders* 140, no. 3 (2012): 205–14.
42. Hidaka, "Depression as a Disease of Modernity," 207.
43. Cited in Ellingson, *The Myth of the Noble Savage*, 91.
44. Ellingson, *The Myth of the Noble Savage*, 66.
45. Cited in Ellingson, *The Myth of the Noble Savage*, 81.
46. Email to author, October 21, 2017.
47. Vargas Llosa, *The Storyteller*, 22.

PART II: RITUAL

1. Chris Hastings, "BBC Turns Its Back on Year of Our Lord: 2,000 Years of Christianity Jettisoned for Politically Correct 'Common Era,'" *Daily Mail*, September 24, 2011, https://www.dailymail.co.uk/news/article-2041265/BBC-turns-year-Our-Lord-2-000-years-Christianity-jettisoned-politically-correct-Common-Era.html.
2. Southern Baptist Convention, "On Retaining The Traditional Method of Calendar Dating (B.C./A.D.)," 2000, http://www.sbc.net/resolutions/298/on-retaining-the-traditional-method-of-calendar-dating--bcad-.
3. Elyse Wanshel, "Pastor Warns Hurricanes Will Hit Cities That Don't Repent 'Sexual Perversion,'" *HuffPost*, September 1, 2017, https://www.huffpost.com/entry/pastor-kevin-swanson-houston-repent-sexual-perversion-hurricane-harvey_n_59a97d74e4b0b5e530fe394d; Tom Barnes, "Indian Health Minister Claims Cancer Is Caused by Sins from a Past Life," *Independent*, November 23, 2017, https://www.independent.co.uk/news/world/asia/india-health-minister-cancer-past-life-sins-assam-state-himanta-biswa-sarma-a8071546.html.

CHAPTER 4: HEY BEAR!

1. Schullery, *Past and Future Yellowstones*, 4.
2. Lafayette Houghton Bunnell, *Discovery of the Yosemite* (New York: Fleming H. Revell Company, 1880), 346.

owstone, 48.

d Richard Erdoes, *Lame Deer, Seeker of Visions* (New York:
2), 130.

Congress, 1st Session, vol. 2 (Washington, DC: Govern-
74), 2107.

'History of Bison Management in Yellowstone," https://
ison-history-yellowstone.htm, accessed September 20, 2019.

and Parks, "Historical Bison Hunt Application Data," http://
anahunt/huntingGuides/bison/bisonFaq.html#question5.

ckfeet to Join Bison Hunt outside Yellowstone," *Bozeman Daily*
, 2018, www.bozemandailychronicle.com/news/environment
son-hunt-outside-yellowstone/article_c780904a-e215-536a
.html.

ss Hunter, 29–30; Wright, "Blackfeet to Join Bison Hunt outside

ittle, "The Buffalo Hunt," *Guardian*, March 7, 2016, https://www
us-news/2016/mar/07/buffalo-hunt-yellowstone-national-park

a Kudelska, "Yakama Nation's First Bison Hunt in Yellowstone,"
Media, January 5, 2018, www.wyomingpublicmedia.org/post
-first-bison-hunt-west-yellowstone#stream/0.

, "An Inside Look at Capturing Yellowstone's Bison: A Photo Essay,"
ch, 11, 2019, http://ricklamplugh.blogspot.com/2018/02/an-inside
ing.html.

ht, "Tribe Responds to Criticism of Elk and Bison Hunt near Yellow-
Park," *Bozeman Daily Chronicle*, March 10, 2017, https://www.bozeman
e.com/news/environment/tribe-responds-to-criticism-of-elk-and-bison
rticle_624d9880-b833-5f9b-9c5f-966b9c2729ec.html.

Nature, 43.

ave the Wild Bison, 225–26.

LET FOOD BE THY MEDICINE

nporary accounts, see Connie Walker and Marnie Luke's extensive cover-
C News, e.g., "'Doctor' Treating First Nations Girls Says Cancer Patients
Themselves,'" November 13, 2014, https://www.cbc.ca/news/indigenous
reating-first-nations-girls-says-cancer-patients-can-heal-themselves
60.

., Åke Hultkrantz, *The Religion of the American Indians*, trans. Monica Setter-
rkeley: University of California Berkeley Press, 1979), 87–89, and Tristan
and Terrol Dew Johnson, "Tohono O'odham *Himdag* and Agri/Culture," in
n and Sustainable Agriculture, ed. Todd LeVasseur et al. (Lexington: University
ntucky Press, 2016), 324.

in Unschuld, *Medicine in China*, 21.

in Ferngren, *Medicine and Religion*, 21.

gren, *Medicine and Religion*, 29.

d in Giorgio Zanchin, "Considerations on 'the Sacred Disease' by Hippocrates,"
rnal of the History of the Neurosciences 1, no. 2 (1992): 92–93.

3. Bunnell, D,
4. Bunnell, Di,
5. Ross-Bryant,
6. Cited in Ross-
7. Pritchard, Pres,
8. Douglas Smith,
   2016): 1.
9. H.G. Wells, A M(
10. Reproduced in Pri\
    Pritchard for much
11. Reproduced in Quan
12. Pritchard, Preserving ,
13. Pritchard, Preserving Ye
14. John Burroughs, "Real a
    298–309. For a detailed a
    Wildlife, Science, and Sentin,
15. Theodore Roosevelt, "'Na
    427–30, reproduced in Ralp
    Temple University Press, 19
16. Cited in Lutts, Nature Fakers,
17. Burroughs, Accepting the Unive,
18. Cited in Pritchard, Preserving Ye
19. Roosevelt, The Wilderness Hunter
    Preserving Yellowstone's Natural Co,
20. Lopez, Of Men and Wolves, 216, 20(
21. W. H. Van Doren, A Suggestive Con,
    Dickinson, 1879), 866.
22. Smith and Ferguson, Decade of the Wo,
23. Described in Bruce Hampton, The Gr(
    1997), 163–64.
24. Peterson, Wolf Nation, 212.
25. Seton, Wild Animals I Have Known, 17–18
26. Seton, Wild Animals I Have Known, 47.
27. Smith and Ferguson, Decade of the Wolf, 88.
28. Smith and Ferguson, Decade of the Wolf, 187.
29. Blakeslee, American Wolf.
30. Quoted in Peterson, Wolf Nation, 188.
31. Blakeslee, American Wolf, 107.
32. Lamplugh, In the Temple of the Wolves, 113.
33. Rick McIntyre, A Society of Wolves: National Parks ,
    ter, MN: Voyageur Press, 1996), 133.
34. From the official Leave No Trace website, https://l\
35. The classic statement on puritanical definitions of wil\
    and the attendant problems when it comes to appreciat
    "The Trouble with Wilderness; or, Getting Back to the
    line: https://www.williamcronon.net/writing/Trouble_wi
36. Cited in Lapham's Quarterly: Book of Nature I, no. 3 (Sum

37. Quoted in Quammen, Yel
38. John (Fire) Lame Deer a
    Simon and Schuster, 197
39. Congressional Record, 43r
    ment Printing Office, 1
40. National Park Service,
    www.nps.gov/articles/
41. Montana Fish, Wildlif
    fwp.mt.gov/hunting/p
42. Michael Wright, "Bl
    Chronicle, February 1
    /blackfeet-to-join-b
    -a539-37432fc3b73
43. Roosevelt, Wilderne
    Yellowstone."
44. Quoted in Joe Wh
    .theguardian.com
    -native-american
45. Quoted in Kami
    Wyoming Publi
    /yakama-nation
46. Rick Lamplugh
    blog post, Mar
    -look-at-captu
47. Michael Wrig
    stone Nationa
    dailychronicl
    -hunt-near/
48. Williams, Il
49. Franke, To

CHAPTER 5
1. For conte
   age for C
   Can Hea
   /doctor-
   -1.2832
2. See, e.g
   wall (B
   Reade
   Religi
   of Ke
3. Cite
4. Cite
5. Fer
6. Cit
   Jou

7. Don Miguel Ruiz, *The Four Agreements* (San Rafael, CA: Amber Allen, 1997), 29.

8. Hippocrates, *On Ancient Medicine*, part 3, trans. Francis Adams, accessed online: http://classics.mit.edu/Hippocrates/ancimed.3.3.html.

9. Information for this section and the citations are from James Opp, *The Lord for the Body: Religion, Medicine, and Protestant Faith Healing in Canada, 1880–1930* (Montreal: McGill-Queen's University Press, 2005), 15–19.

10. Cited in Whorton, *Nature Cures*, 6.

11. Cited frequently, e.g., W. F. Bynum, "Clinical Precision," *Nature* (March 24, 2011).

12. Bigelow, *A Discourse on Self-Limited Diseases*, 34.

13. Timothy Childs, *Rational Medicine* (New York: Baker and Godwin, 1863), 20.

14. Cited in Gene Fowler, *Mystic Healers and Medicine Shows* (Santa Fe: Ancient City Press, 1997), 120.

15. Joseph E. Meyer, *Catalog* (Hammond, IN: Indiana Botanic Gardens, 1933).

16. Graham, *Lectures on the Science of Human Life*, 72–73. In the same section, Graham takes care to distinguish "natural," which he approves of, from "savage," which he does not.

17. *Good Health* 2, no. 7 (July 1, 1907): 90, accessed online: https://archive.org/details/good-health-volume-42-issue-07-july-1st-1907/page/n89.

18. *The Medical and Surgical Reporter: A Weekly Journal*, vol. 3 (Philadelphia: Crissy & Markley, 1860), 516.

19. Joseph Meyer, *Nature's Remedies* (Hammond, IN: Indiana Botanic Gardens, 1834), 64.

20. Ehrenreich, *Natural Causes*, 14–15.

21. Statement published in *Two Row Times*, November 4, 2014, https://tworowtimes.com/news/exclusive-family-six-nations-girl-using-indigenous-medicine-issues-statement.

22. Hamilton Health Sciences Corp. v. D.H., 2014 ONCJ 603, p. 77.

23. *Hamilton Health Sciences Corp. v. D.H.*, 77.

24. Cited in *Hamilton Health Sciences Corp. v. D.H.*, 78.

25. Ted J. Kaptchuk and David M. Eisenberg, "The Persuasive Appeal of Alternative Medicine," *Annals of Internal Medicine* 129, no. 12 (1998): 1063.

26. Colleen Derkatch, "The Self-Generating Language of Wellness and Natural Health," *Rhetoric of Health & Medicine* 1, no. 1 (2018): 132–60.

27. Anna Maria Clement, "From Agony to Ecstasy," *Healing Our World* 37, no. 4 (2017): 20.

28. Claudia Kalb, "Faith and Healing," *Newsweek*, November 9, 2003, https://www.newsweek.com/faith-healing-133365.

29. Roger S. Ulrich, "Effects of Healthcare Environmental Design on Medical Outcomes," *Design and Health: Proceedings of the Second International Conference on Health and Design. Stockholm, Sweden: Svensk Byggtjanst* 49 (2001): 40.

30. George Hadjipavlou et al., "'All My Relations': Experiences and Perceptions of Indigenous Patients Connecting with Indigenous Elders in an Inner City Primary Care Partnership for Mental Health and Well-Being," *Canadian Medical Association Journal* 190, no. 20 (2018): E608–E615.

31. *British Medical Journal*, "Too Much Medicine," https://www.bmj.com/too-much-medicine.

32. Reich, *Calling the Shots*, 117.

33. Ehrenreich, *Natural Causes*, 209.

CHAPTER 6: DEEPAK CHOPRA'S CONDO

1. Robin Finn, "Health-Centric Homes, for a Price," *New York Times*, June 28, 2013.
2. See the Goop website: https://shop.goop.com/shop/products/eau-de-parfum-edition
   -01-church?country=USA.
3. Quoted in Taffy Brodesser-Akner, "How Goop's Haters Made Gwyneth Paltrow's
   Company Worth $250 Million," *New York Times Magazine*, July 25, 2018, https://
   www.nytimes.com/2018/07/25/magazine/big-business-gwyneth-paltrow-wellness.html.
4. "Clean Beauty and Why It's Important," Goop, https://goop.com/beauty/personal
   -care/clean-beauty-and-why-its-important.
5. "Premium Body Care Standards," Whole Foods Market, https://www.wholefoods
   market.com/about-our-products/premium-body-care-standards. accessed September 20, 2019.
6. Laudan, *Cuisine and Empire*, 44.
7. Cederström and Spicer, *The Wellness Syndrome*, 49.
8. George Orwell, *The Road to Wigan Pier* (Mariner Books, 1972), 127–28.
9. Jennifer Gunter, "In Goop Health: Wellness Panem Style," January 28, 2018,
   https://drjengunter.com/2018/01/28/in-goop-health-wellness-panem-style.
10. For quotes, see the official website, https://tourmalinespring.com, accessed Septem-
    ber 20, 2019.
11. Prices at the bottom of the following website's page: https://chopra.com/live-events
    /6-day-perfect-health.
12. Douglas, *Purity and Danger*, 37.
13. Quotes from the book's website available at http://www.amberallen.com/product
    /books/the-seven-spiritual-laws-of-success, accessed September 20, 2019.
14. Joel Osteen, *Your Best Life Now* (New York: FaithWords, 2015), 256.
15. Quoted in Howard Snyder, *Populist Saints: B. T. and Ellen Roberts and the First Free
    Methodists* (Grand Rapids, MI: Eerdmans, 2006), 250.
16. Snyder, *Populist Saints*, 550.
17. Eloise Blondiau, "Who Deserves to Be Healthy? The Prosperity Gospel According
    to Goop," *America*, June 7, 2018, https://www.americamagazine.org/arts-culture
    /2018/06/07/who-deserves-be-healthy-prosperity-gospel-according-goop.
18. David Roberts, "Wealthier People Produce More Carbon Pollution—Even the Green
    Ones," *Vox*, December 1, 2017, https://www.vox.com/energy-and-environment/2017
    /12/1/16718844/green-consumers-climate-change.
19. Quoted by Zack Slobig of the Skoll Foundation on Twitter, from a speech given by
    Xavier Berrara at the Global Climate Action Summit, September 12–14, 2018,
    https://twitter.com/slobig/status/1040638549526106114?s=20.
20. James Hamblin, "The Seduction of Wellness Real Estate," *Atlantic*, February 27,
    2017, https://www.theatlantic.com/health/archive/2017/02/wellness-real-estate
    /517560.
21. Simona Fischer, "The WELL Building Standard: Not to Be Used Alone," *Building
    Green*, March 8, 2017, https://www.buildinggreen.com/op-ed/well-building
    -standard-not-be-used-alone.
22. Emily Atkin, "Do You Know Where Your Healing Crystals Come From?," *New
    Republic*, May 11, 2018, https://newrepublic.com/article/148190/
    know-healing-crystals-come-from.

PART III: LAW

1. For the citations in this section, see Drummond, *Natural Law in the Spiritual World*, 317–37.

2. National Academy of Sciences, *Science and Creationism* (Washington, DC: National Academies Press, 1999), xi.

3. Burlamaqui, *The Principles of Natural and Politic Law*, 4. See 11–18 for a tortured explanation of why humans in the state of nature were not following natural law, but how, if they had been, everything would have been perfect.

4. See John Finnis's defense of natural law theories against the "appeal to nature fallacy" accusation in "Natural Law Theories," section 1.1.1, *Stanford Encyclopedia of Philosophy*, https://plato.stanford.edu/entries/natural-law-theories/#BasReaForAct NeeForGovAut. Finnis rebuts the idea that natural law has anything to do with "nature" traditionally understood, instead defining it as "the law of reason" and tracing this sense back to Aquinas. What Finnis does not do, however, is contrast the supposed meaning of "natural" in natural law with the tendency of philosophers, theologians, and politicians to see moral laws in the laws of the natural world. If supporters of natural law do not commit the "appeal to nature fallacy" in theory, history demonstrates that in practice they certainly do. Political thinkers tie themselves in knots trying to harmonize various senses of nature, lurching between one sense of the word and another, usually with circular reasoning that serves to confirm the assumptions of the thinker. As Carl Becker remarks wryly, it is difficult to squeeze arguments out of the "endless, half-deciphered Book of Nature," and what results is (in the example Becker gives of John Locke) "lumbering, involved, obscured by innumerable and conflicting qualifications—a dreary devil of an argument." Becker, *The Declaration of Independence*, 58, 72. See 24–79 for his discussion of the various senses of "nature," including that of the natural world in "natural rights" philosophy.

5. Constantin François de Chassebœuf, *Volney's Ruins* (New York: Vale, 1853), 185. See 185–97 for his entire treatment of nature's laws, the inconsistencies of which give insight into the problems faced by attributing all regularities in nature to a God concerned with the benefit of mankind. Thus "it is a law of nature that water flows downwards" and that "it is heavier than air," and these laws have an "express clause of punishment attending to the infraction of them."

6. Chassebœuf, *Volney's Ruins*, 189.

7. Noel Antoine Pluche, *Nature Delineated: Vol II*, trans. John Kelly et al. (London: James Hodges, 1740), 191. The many volumes of Pluche's histories include a number of divine just-so morality stories, as well as beautiful fold-out woodcuts and a dialogue format that made the information more appealing and didactic.

8. "Sen. Sasse on the Rise of 'Anti-tribes' and a Growing American Tolerance for Lies," *PBS NewsHour*, October 23, 2018; full interview transcript: https://www.pbs .org/newshour/show/sen-sasse-on-the-rise-of-anti-tribes-and-a-growing-american -tolerance-for-lies.

9. Quoted in Sarah Taylor, "Televangelist on Hurricane Harvey: 'Flood Is from God,' a 'Judgment on America,'" *Blaze*, September 7, 2017, https://www.theblaze.com /news/2017/09/07/televangelist-jim-bakker-on-hurricane-harvey-flood-is-from -god-a-judgment-on-america.

10. Quoted in Graeme Wood, "His Kampf," *Atlantic*, June 2017, https://www.theatlantic
.com/magazine/archive/2017/06/his-kampf/524505.

CHAPTER 7: THE INVISIBLE HAND

1. Zhill Olonan and Geoffrey Woo, "From Keto to Carnivore: Did Humans Evolve
from a Meat-Based Diet? ft. Travis Statham," October 31, 2018; podcast transcription
at https://hvmn.com/podcast/from-keto-to-carnivore-ft-travis-statham-episode-90.

2. "Episode 28: Michael Goldstein & Saif Ammous," *Human Performance Outliers Podcast*, August 1, 2018, https://humanperformanceoutliers.libsyn.com/episode-28
-michael-goldstein-saif-ammous.

3. Jordan Pearson, "Inside the World of the 'Bitcoin Carnivores,'" *Vice*, September 29,
2017, https://www.vice.com/en_us/article/ne74nw/inside-the-world-of-the-bitcoin
-carnivores.

4. Nassim Taleb, foreword to Saifedean Ammous, *The Bitcoin Standard: The Decentralized Alternative to Central Banking* (Hoboken, NJ: John Wiley & Sons, 2018), xiv.

5. Arthur Levitt, "The National Market System: A Vision That Endures," US Securities and Exchange Commission, January 8, 2001, https://www.sec.gov/news/speech
/spch453.htm. To emphasize the connection between economics and nature, Levitt
begins the speech with a parallel between the "force of the natural world" and the
"formidable drive of the entrepreneurial spirit."

6. Team Luno, "The Natural Evolution of Money," Luno, https://www.luno.com/blog
/en/post/the-natural-evolution-of-money, October 14, 2018.

7. Ammous, *The Bitcoin Standard*, 35.

8. There's a great deal of controversy in philosophy of science over the nature of models
and how to measure their utility and relative correspondence to truth. With regard to
economics specifically, see e.g., Mäki, *Fact and Fiction in Economics*. West's *Scale* makes
a fascinating case for mathematical laws that govern growth across natural and cultural systems. For a rigorous attempt at using evolutionary theory to model economics, see, e.g., Robert H. Frank. *The Darwin Economy: Liberty, Competition, and the
Common Good* (Princeton, NJ: Princeton University Press, 2011). The advent of
quantum mechanics ushered in another tempting natural system for modeling economics, e.g., David Orrell, *Quantum Economics: The New Science of Money* (London:
Icon Books, 2018). There are skeptics, however, who discount even the most sophisticated and responsible attempts at this sort of modeling. See Mirowski's polemical
More Heat than Light and his edited volume, *Natural Images in Economic Thought*.

9. Plato, *Republic*, 370a–b, trans. Paul Shorey (Cambridge, MA: Harvard University
Press, 1969), accessed through the Tufts online classics collection: http://www
.perseus.tufts.edu/hopper/text?doc=Perseus%3Atext%3A1999.01.0168%3Abook
%3D2%3Asection%3D370a.

10. Aristotle, *Politics*, part V, trans. Benjamin Jowett, accessed through the MIT online
classics collection: http://classics.mit.edu/Aristotle/politics.1.one.html. All quotes
come from this section.

11. Plato, *Republic*, 462c–d.

12. Aristotle, *Politics*, part V.

13. Adam Smith, *An Inquiry into the Nature and Causes of the Wealth of Nations by Adam
Smith, Edited with an Introduction, Notes, Marginal Summary and an Enlarged Index by*

Edwin Cannan, vol. 2 (1776; London: Methuen, 1904), https://oll.libertyfund.org
/titles/119#Smith_0206-02_312.

14. Smith, *Wealth of Nations*, vol. 1, https://oll.libertyfund.org/titles/237#Smith_0206-01
_1016.

15. Citations in this paragraph from Lisa Hill, "The Hidden Theology of Adam Smith,"
European Journal of the History of Economic Thought 8, no. 1 (2001): 1–29.

16. Burke, *Thoughts and Details on Scarcity*, 32. Despite appealing to nature in this instance, Burke was well aware that certain understandings of naturalness worked
against his preferred forms of government and religion. Some aspects of nature, he
acknowledged, were in need of improvement. His satirical book, *A Vindication of
Natural Society: or, a View of the Miseries and Evils Arising to Mankind from Every Species of Artificial Society*, attempted to defend revealed religion over "natural" religion
by demonstrating the absurdity of support for a purely natural society.

17. Jean-Jacques Rousseau, "Discourse on Political Economy," *The Basic Political Writings*, trans. Donald Cress (Indianapolis: Hackett, 2011), 125–26.

18. I. Bernard Cohen, "Newton and the Social Sciences, with Special Reference to Economics, or, the Case of the Missing Paradigm," 55–90, in Mirowski ed., *Natural Images in Economic Thought*.

19. Ralph Waldo Emerson, "Wealth," *The Works of Ralph Waldo Emerson*, vol. 2 (London: George Bell and Sons, 1904), 245.

20. Quoted in Eve Gerber, "The Best Books on Globalization," Fivebooks.com, 2017,
https://fivebooks.com/best-books/larry-summers-globalization.

21. Henry Linville, *The Biology of Man and Other Organisms* (New York: Harcourt, Brace,
1923), 168–77. Counterintuitively, the textbook lambasts "strong members of our society" who "control the energies of their fellow men." It holds out hope that, through
eugenics, we can eventually eliminate those who "grab the lion's share of every commercial or industrial opportunity." Fitness is not the equivalent of morality.

22. Andrew Carnegie, *The Autobiography of Andrew Carnegie and His Essay the Gospel of
Wealth* (1920; Mineola, NY: Dover Thrift Editions, 2014), 249–50.

23. Cited in Alexander Rosenberg, "Does Economic Theory Give Inspiration to Economics?," in Mirowski, *Natural Images in Economic Thought*, 390.

24. Buckley, *Up from Liberalism*, 202.

25. For a reappraisal of Hofstadter's *Social Darwinism in American Thought*, and a comprehensive look at related issues, see Kaye, *Social Meaning of Modern Biology*.

26. For an excellent comparison of Paine and Burke on the meaning of "nature," see
Levin, *The Great Debate*, 43–69.

27. Thomas Paine, "Rights of Man," *Rights of Man, Common Sense, and Other Political
Writings* (1791; Oxford: Oxford University Press, 1998), 135.

28. Gautier, *Le Darwinisme Social*, 6. Thanks to Daniel Morgan for the translation of
this and the following passages.

29. Gautier, *Le Darwinisme Social*, 15, 53.

30. Peter Kropotkin, *Mutual Aid: A Factor of Evolution* (New York: McClure Phillips,
1902), ix.

31. James Pusey, "Global Darwin: Revolutionary Road," *Nature* 462, no. 7270 (2009):
162–63, https://www.nature.com/articles/462162a.

32. Pusey, "Global Darwin."

33. David Flannery, "Global Darwin: Ideas Blurred in Early Eastern Translations," *Nature* 464, no. 984 (2009): https://www.nature.com/articles/462984c.
34. Cited in Pusey, *China and Charles Darwin*, 452.
35. Joseph Stalin, *Dialectical and Historical Materialism* (New York: International Publishers, 1940), 9, 13. See 5–16 for a full (if confused) account of his understanding of nature and its relationship to history, politics, and philosophy.
36. Kaye, *The Social Meaning of Modern Biology*, 29.
37. Darwin, *Origin of Species*, 489.
38. Quoted in Hofstadter, *Social Darwinism*, 85.
39. Oscar Schmidt, "Science and Socialism," *Popular Science Monthly* 14, no. 5 (1879): 577–91.
40. From the website InvestingAnswers, https://investinganswers.com/dictionary/c/collusion, accessed September 20, 2019.
41. Cited in Rocke, *Forbidden Friendships*, 3.
42. Aristotle, *Politics*, part X. Here Aristotle commits a complicated version of the appeal to nature fallacy by defining the "essence" of money according to a "natural kind" found in the natural world. Defenders of natural law argue that Aristotle, and later Aquinas, only mean "natural" as "essence" when they talk about something being natural. However, by importing the idea of breeding from the natural world, Aristotle demonstrates that, in fact, his understanding of money's "nature" depends on a metaphor—*money as breeding creature*—that assumes there are moral lessons in the order of nature.
43. Christopher Jelinger, *Usury Stated Overthrown* (London: J. Wright, 1679), 14.

CHAPTER 8: THE RHYTHM

1. Joshua J. McElwee, "Francis Lambasts International Aid, Suggests Catholics Should Limit Children," *National Catholic Reporter*, January 19, 2015, https://www.ncronline.org/news/francis-lambasts-international-aid-suggests-catholics-should-limit-children.
2. There are many different terms for birth control methods that don't depend on pharmaceuticals, and specialists refer to them under the umbrella term of "fertility awareness based methods." Chelsea Polis, an expert on FABMs, told me that "natural family planning" generally has religious connotations, and requires abstinence during the "fertile window," whereas other FABMs might allow for sexual activity during the fertile window (with a condom, say). See also Chelsea B. Polis and Rachel K. Jones, "Multiple Contraceptive Method Use and Prevalence of Fertility Awareness Based Method Use in the United States, 2013–2015," *Contraception* 98, no. 3 (2018): 188–92.
3. Quoted in Tone, *Devices and Desires*, 44.
4. Joseph Reiner, introduction to Latz, *The Rhythm of Sterility and Fertility in Women*, 1.
5. Latz, *The Rhythm of Sterility and Fertility in Women*, 103.
6. Pope Paul VI, *Humanae vitae*, sections 15 and 16: http://w2.vatican.va/content/paul-vi/en/encyclicals/documents/hf_p-vi_enc_25071968_humanae-vitae.html.
7. Quoted in Malcolm Gladwell, "John Rock's Error," in *What the Dog Saw: And Other Adventures* (New York: Little, Brown, 2009), 105.
8. Gladwell, "John Rock's Error," 106.
9. Gladwell, "John Rock's Error," 107.
10. Ezra Heywood, *Cupid's Yokes: Or, the Binding Forces of Conjugal Life* (Princeton, NJ: Co-operative Publishing, 1877).

11. Gordon, *The Moral Property of Women*, 57.
12. Cited in Jütte, *Contraception*, 214.
13. See May, *America and the Pill*, 131–32.
14. Grigg-Spall, *Sweetening the Pill*, 189.
15. Carolyn Moynihan, "Sweetening the Pill," *Mercatornet*, October 15, 2013, https://www.mercatornet.com/articles/view/sweetening_the_pill/12928.
16. Latz, *The Rhythm of Sterility and Fertility in Women*, 114.
17. Latz, *The Rhythm of Sterility and Fertility in Women*, 114.
18. Quoted in Tentler, *Catholics and Contraception*, 120.
19. Quoted in Tentler, *Catholics and Contraception*, 121.
20. Quoted in Stone, *Sex and the Constitution*, 198.
21. Latz, *The Rhythm of Sterility and Fertility in Women*, 120.
22. In this passage Blackstone must remain, for obvious reasons, somewhat ambiguous, and asserts that the unspeakable act (which is worse than rape) can be committed "with man or beast." William Blackstone, *Commentary on the Laws of England* (Oxford: Clarendon Press, 1765–69), book 4, chapter 15: https://ebooks.adelaide.edu.au/b/blackstone/william/comment/book4.15.html.
23. Cited in Jütte, *Contraception*, 25, 76.
24. Theo van der Meer, "'Are Those People Like Us'—Early Modern Homosexuality in Holland," in *Queer Masculinities: 1550–1800: Siting Same-Sex Desire in the Early Modern World*, ed. Katherine O'Donnell and Michael O'Rourke (London: Palgrave, 2006), 58.
25. See David M. Kennedy, *Birth Control in America: The Career of Margaret Sanger* (New Haven, CT: Yale University Press, 1971), 45; Mark Van Doren, "Comstock, Anthony," in *Dictionary of American Biography*, ed. Allen Johnson and Dumas Malone (New York: Charles Scribner's Sons, 1930), 330–31.
26. Massachusetts General Laws, Part IV, Title I, Chapter 272, Section 34 (Chapter 272 is titled "Crimes against Chastity, Morality, Decency and Good Order"), https://malegislature.gov/laws/generallaws/partiv/titlei/chapter272/section34, accessed September 20, 2019.
27. Carol Flora Brooks, "The Early History of the Anti-Contraceptive Laws in Massachusetts and Connecticut," *American Quarterly* 18, no. 1 (Spring 1966): 3.
28. Cited in Noonan, *Contraception*, 446. See 438–76 for a detailed discussion of related issues.
29. Cited in Bernard Doering, "On Good Authority?," *Commonweal*, March 12, 2012, https://www.commonwealmagazine.org/good-authority.
30. Trent Horn, "Why the Church Cannot Marry the Impotent," *Catholic Answers*, August 7, 2014, https://www.catholic.com/magazine/online-edition/why-the-church-cannot-marry-the-impotent.
31. Gregory K. Popcak, *Holy Sex: A Catholic Guide to Toe-Curling, Mind-Blowing, Infallible Loving* (New York: Crossroad Publishing, 2008), 248.
32. Popcak, *Holy Sex*, 136.
33. Quoted in Wills, *Papal Sin*, 90.
34. Wills, *Papal Sin*, 90.
35. Wills, *Papal Sin*, 91.
36. The classic statement of the argument in favor of heeding disgust is bioethicist Leon Kass's "The Wisdom of Repugnance," *New Republic*, June 2, 1997. After some

throat-clearing he lays out the case: "Repugnance is the emotional expression of deep wisdom, beyond reason's power to articulate it." Kass is arguing in this context against human cloning—the article comes complete with a picture of Frankenstein's monster.

37. James R. Spence, "The Law of Crime against Nature," *North Carolina Law Review* 32, no. 3 (1954): 321–22.

38. Ida B. Wells, *The Light of Truth: Writings of an Anti-Lynching Crusader* (New York: Penguin, 2014). See chapter 2 for the quote and full account.

39. Covered in Jonathan Capehart, "How the Terror of Lynchings in the Past Haunts Us Today and Our Future," *Washington Post*, June 27, 2017. https://www.washington post.com/blogs/post-partisan/wp/2017/06/27/how-the-terror-of-lynchings-in-the -past-haunt-us-today-and-our-future/?utm_term=.857e7f949e9b.

40. Quoted in Randall Kenny, *Interracial Intimacies: Sex, Marriage, Identity, and Adoption* (New York: Vintage Books, 2003), 84.

41. Cited in Pascoe, *What Comes Naturally*, 70–71.

42. Pascoe, *What Comes Naturally*, 61, 71.

43. Pascoe, *What Comes Naturally*, 188.

44. Pascoe, *What Comes Naturally*, 189.

45. For example, Pope Pius XII, "Discrimination and Christian Conscience," in *Pastoral Letters of the United States Catholic Bishops*, vol. 2, *1941–61*, ed. Hugh J. Nolan (Washington, DC: National Conference of Catholic Bishops, United States Catholic Conference, 1983), 201–6. Although for a more nuanced take on interracial marriage and the Church, see Pascoe, *What Comes Naturally*, 209–12.

46. Pascoe, *What Comes Naturally*, 191.

47. RNC Communications, "The 2016 Republican Party Platform," July 2016, https:// www.gop.com/the-2016-republican-party-platform, accessed September 20, 2019.

48. Planned Parenthood, "What Causes Sexual Orientation?," https://www.planned parenthood.org/learn/sexual-orientation-gender/sexual-orientation/what-causes -sexual-orientation, accessed September 20, 2019.

49. Covered in Zoë Schlanger, "The Gulls Are Alright: How a Lesbian Seagull Controversy Shook Up 1970s Conservatives," *Quartz*, July 10, 2017, https://qz.com/1023638 /the-gulls-are-alright-how-a-lesbian-seagull-discovery-shook-up-1970s-conservatives.

50. Quoted in Glen Mills, "A Look Back at the 42-Year Career of Senator Orrin Hatch," ABC4 News, November 7, 2018, https://www.abc4.com/news/local-news /a-look-back-at-the-42-year-career-of-senator-orrin-hatch.

51. Ryan and Jethá, *Sex at Dawn*, 1.

52. For an excellent survey, see Melissa Hogenboom, "Are There Any Homosexual Animals?," BBC, February 6, 2015, http://www.bbc.com/earth/story/20150206-are -there-any-homosexual-animals. Hogenboom points out that the only other animal in which we regularly see sustained same-sex preference and coupling is domestic sheep.

CHAPTER 9: GOD-GIVEN TALENT

1. See Dave Pearce's answer here: https://www.quora.com/Is-it-possible-to-build-a -Ronnie-Coleman-body-naturally-without-using-steroids#, accessed September 20, 2019.

2. Simon Turnbull, "Maurice Greene: 'You Can Do This Clean and with Just God Given Talent,'" *Independent*, June 5, 2005, https://www.independent.co.uk/sport /general/maurice-greene-you-can-do-this-clean-and-with-just-god-given-talent -493664.html.

3. *Congressional Record*, 108th Congress, 2nd session, vol. 150, www.govinfo.gov/content /pkg/CREC-2004-06-02/pdf/CREC-2004-06-02.pdf.

4. Julian Savulescu, Bennett Foddy, and Megan Clayton, "Why We Should Allow Performance Enhancing Drugs in Sport," *British Journal of Sports Medicine* 38, no. 6 (2004): 666–70.

5. "The Pros and Cons of Cortisone," *Golf Digest*, September 19, 2011, https://www .golfdigest.com/story/kaspriske-fitness-column-2011-11.

6. See Karen Abbott, "The 1904 Olympic Marathon May Have Been the Strangest Ever," *Smithsonian*, August 7, 2012, https://www.smithsonianmag.com/history/the -1904-olympic-marathon-may-have-been-the-strangest-ever-14910747, and P. McCrory, "Last Orders, Gents . . . ," *British Journal of Sports Medicine* 39 (2005): 879, https://bjsm.bmj.com/content/bjsports/39/12/879.1.full.pdf.

7. Roger Pielke Jr. and Ross Tucker, "Investigation of Cross-US Run Attempt by Rob Young," September 2016, accessed at https://www.letsrun.com/wp-content/uploads /2016/10/Robert-Young-Skins-Report-Cheat.pdf.

8. IAAF, *Competition Rules 2018–2019* (Monaco: IAAF, 2017), 73, available at https:// www.iaaf.org/about-iaaf/documents/rules-regulations.

9. IAAF, "IAAF Council Introduces Rule regarding 'Technical Aids,'" March 26, 2007, https://www.iaaf.org/news/news/iaaf-council-introduces-rule-regarding-techni.

10. Some material and citations in the next three paragraphs adopted from Alan Levinovitz, "In an Era of Doping and Blade-Running, What Is a 'Natural Athlete,' Anyway?," *Washington Post*, August 1, 2016, https://www.washingtonpost.com/national /health-science/in-an-era-of-doping-and-blade-running-what-is-a-natural-athlete -anyway/2016/08/01/a675e3e2-42b6-11e6-88d0-6adee48be8bc_story.html.

11. IAAF, "Constitution," November 1, 2017, 7: https://www.iaaf.org/about-iaaf /documents/constitution.

12. Quoted in Roger Abrams, *Playing Tough: The World of Sports and Politics* (Lebanon, NH: Northeastern University Press, 2013), 15.

13. Quoted in Colleen English, "'Not a Very Edifying Spectacle': The Controversial Women's 800-Meter Race in the 1928 Olympics," *Sport in American History*, October 8, 2015, https://ussporthistory.com/2015/10/08/not-a-very-edifying-spectacle -the-controversial-womens-800-meter-race-in-the-1928-olympics.

14. See Roger Robinson's thorough investigation, "Eleven Wretched Women," *Runner's World*, May 14, 2012, https://www.runnersworld.com/advanced/a20802639/eleven -wretched-women.

15. Quoted in Sydney Pereira, "Women at the Winter Olympics Always Had Fewer Chances, and Science Says That's Bunk," *Newsweek*, February 18, 2018, https:// www.newsweek.com/women-winter-olympics-have-always-had-fewer-chances -compete-and-science-says-808400.

16. Roslyn Kerr, "Why It Might Be Time to Eradicate Sex Segregation in Sports," *The Conversation*, January 14, 2018, https://theconversation.com/why-it-might-be-time -to-eradicate-sex-segregation-in-sports-89305.

17. Ross Tucker, "High Testosterone, Unfair Advantages and Principles of Transparency," commentary posted on *The Science of Sport*, August 2, 2018, https://sports scientists.com/2018/08/letter-to-bjsm-reinforcing-call-for-retraction-of-iaaf -research-on-testosterone-in-women.

18. Valérie Thibault et al., "Women and Men in Sport Performance: The Gender Gap Has Not Evolved since 1983," *Journal of Sports Science & Medicine* 9, no. 2 (2010): 214–23.

19. María José Martínez-Patiño, "Personal Account: A Woman Tried and Tested," *Lancet* 366 (2005): S38.

20. Martínez-Patiño, "Personal Account," S38.

21. Quoted in Christopher Clarey and Gina Kolata, "Gold Awarded amid Dispute over Runner's Sex," *New York Times*, August 20, 2009, https://www.nytimes. com/2009/08/21/sports/21runner.html.

22. Tom Bryant, "Caster Semenya Subjected to 'Humiliating' Sex Test, Claims Coach," *Guardian*, September 17, 2009, https://www.theguardian.com/sport/2009/sep/17 /caster-semenya-sex-test-athletics.

23. Quoted in PTI, "Dutee Failed Test Conducted to Check Androgen Level: SAI," *Times of India*, updated July 16, 2014, https://timesofindia.indiatimes.com/sports /more-sports/athletics/Dutee-failed-test-conducted-to-check-androgen-level-SAI /articleshow/38495489.cms.

24. CAS 2014/A/3759 Dutee Chand v. Athletics Federation of India & the International Association of Athletics Federations, 32, 8.

25. Jeré Longman, "Caster Semenya Will Challenge Testosterone Rule in Court," *New York Times*, June 18, 2018, https://www.nytimes.com/2018/06/18/sports/caster -semenya-iaaf-lawsuit.html.

26. IAAF, "Eligibility Regulations for the Female Classification (Athletes with Differences of Sex Development) Explanatory Notes/Q&A," April 2018, 6, https://www .sportsintegrityinitiative.com/wp-content/uploads/2018/04/Explanatory-Notes _-IAAF-Eligibility-Regulations-for-the-Female-Classification.pdf.

27. Pielke, *The Edge*, 195.

28. Katrina Karkazis et al., "Out of Bounds? A Critique of the New Policies on Hyperandrogenism in Elite Female Athletes," *American Journal of Bioethics* 12, no. 7 (2012): 3–16, 14.

29. Joanna Harper, "Race Times for Transgender Athletes," *Journal of Sporting Cultures and Identities* 6, no. 1 (2015): 1–9.

30. Shasta Darlington, "Transgender Volleyball Star in Brazil Eyes Olympics and Stirs Debate," *New York Times*, March 17, 2018, https://www.nytimes.com/2018/03/17 /world/americas/brazil-transgender-volleyball-tifanny-abreu.html.

AFTERWORD: SALVATION

1. Howard, *An Agricultural Testament*, 1.

2. Howard, *Agricultural Testament*, 37, cited in Mann, *The Wizard and the Prophet*, 182.

3. Mann, *The Wizard and the Prophet*, 179.

4. Based on Arthur Basham's translation, *The Wonder That Was India* (London: Sidgwick and Jackson, 1954), 247–48. Some elements have been modified following Wendy Doniger's translation in *The Rig Veda* (New York: Penguin Classics, 2005), 25–26, and Joel P. Brereton's translation in "Edifying Puzzlement: Ṛgveda 10.129 and the Uses of Enigma," *Journal of the American Oriental Society* (1999): 248–60.

Select Bibliography

Allenby, Braden R., and Daniel Sarewitz. *The Techno-Human Condition*. Cambridge, MA: MIT Press, 2011.

Anderson, E. H. *Caring for Place: Ecology, Ideology, and Emotion in Traditional Landscape Management*. Walnut Creek, CA: Left Coast Press, 2014.

Bailey, L. H. *The Holy Earth*. New York: Charles Scribner's Sons, 1915.

Ball, Philip. *Unnatural: The Heretical Idea of Making People*. London: Bodley Head, 2011.

Becker, Carl. *The Declaration of Independence: A Study in the History of Political Ideas*. New York: Vintage Books, 1942.

Berry, Wendell. *The Unsettling of America: Culture and Agriculture*. San Francisco: Sierra Club Books, 1978.

Bigelow, Jacob. *A Discourse on Self-Limited Diseases*. Boston: Nathan Hale, 1835.

Biss, Eula. *On Immunity*. Minneapolis: Graywolf Press, 2014.

Blakeslee, Nate. *American Wolf: A True Story of Survival and Obsession in the West*. New York: Crown, 2017.

Bradley, Robert. *Husband-Coached Childbirth: The Bradley Method of Natural Childbirth*. 1965. Reprint, New York: Bantam, 2008.

Buckley, William F., Jr. *Up from Liberalism*. New York: McDowell, 1959.

Burke, Edmund. *Thoughts and Details on Scarcity*. London: F. C. Rivington, 1800.

Burlamaqui, J. J. *The Principles of Natural and Politic Law*, Vol. II. London: J. Nourse, 1763.

Burroughs, John. *Accepting the Universe*. Boston: Houghton Mifflin, 1920.

Bush, Barbara. *Slave Women in Caribbean Society, 1650–1838*. Indianapolis: Indiana University Press, 1990.

Butler, Samuel. *Erewhon: Or, Over the Range*. London: Jonathan Cape, 1923.

Cameron, Catherine. *Captives: How Stolen People Changed the World*. Lincoln: University of Nebraska Press, 2016.

Cameron, Ken. *Vanilla Orchids: Natural History and Cultivation*. Portland: Timber Press, 2012.

Carson, Rachel. *Silent Spring*. Boston: Houghton Mifflin, 1962.

Cederström, Carl, and André Spicer. *The Wellness Syndrome*. Cambridge, UK: Polity Press, 2015.

Chase, Stuart. *Men and Machines*. New York: Macmillan, 1929.

Conrad, Joseph. *Heart of Darkness*. 1899. Reprint, Portland: Tin House Books, 2013.

Darwin, Charles. *The Works of Charles Darwin*, vol. 10, *The Foundations of the Origin of Species*. Edited by Paul H. Barrett and R. B. Freeman. London: Routledge, 2016.

———. *On the Origin of Species*. London: John Murray, 1859.

Dekker, Sidney. *The End of Heaven: Disaster and Suffering in a Scientific Age*. New York: Routledge, 2017.

Descola, Phillipe. *Beyond Nature and Culture*. Chicago: University of Chicago Press, 2013.

Dillard, Annie. *Pilgrim at Tinker Creek*. New York: Harper's Magazine Press, 1974.

Douglas, Mary. *Purity and Danger: An Analysis of Concepts of Pollution and Taboo*. 1966. Reprint, New York: Routledge Classics, 2009.

Drummond, Henry. *Natural Law in the Spiritual World*. London: Hodder and Stoughton, 1910.

Ecott, Tim. *Vanilla: Travels in Search of the Ice Cream Orchid*. New York: Atlantic Monthly Press, 2005.

Ehrenreich, Barbara. *Natural Causes: An Epidemic of Wellness, the Certainty of Dying, and Killing Ourselves to Live Longer*. New York: Hachette, 2018.

Eliade, Mircea. *The Myth of the Eternal Return: Cosmos and History*. New York: Pantheon Books, 1955.

Ellingson, Ter. *The Myth of the Noble Savage*. Berkeley: University of California Press, 2001.

Evans, Bergen. *The Natural History of Nonsense*. New York: Knopf, 1946.

Evelyn, John. *Fumifugium*. 1661. Reprint, Oxford, UK: Old Ashmolean Reprints, 1930.

Fadiman, Anne. *The Spirit Catches You and You Fall Down: A Hmong Child, Her American Doctors, and the Collision of Two Cultures*. 1997. Reprint, New York: Farrar, Straus, and Giroux, 2012.

Ferngren, Gary B. *Medicine and Religion: A Historical Introduction*. Baltimore: Johns Hopkins University Press, 2014.

Franke, Mary Ann. *To Save the Wild Bison: Life on the Edge in Yellowstone*. Norman: University of Oklahoma Press, 2005.

Gautier, Émile. *Le Darwinisme Social*. Paris: Derveaux, 1880.

Gordon, Linda. *The Moral Property of Women: A History of Birth Control Politics in America*. Chicago: University of Illinois Press, 2002.

Graham, Sylvester. *A Treatise on Bread, and Bread-Making*. Boston: Light & Stearns, 1837.

———. *Lectures on the Science of Human Life*. Battle Creek, MI: Office of the Health Reformer, 1872.

Grigg-Spall, Holly. *Sweetening the Pill: Or How We Got Hooked on Hormonal Birth Control*. Alresford, UK: Zero Books, 2013.

Hadot, Pierre. *The Veil of Isis: An Essay on the History of the Idea of Nature*. Translated by Michael Chase. Cambridge, MA: Harvard University Press, 2006.

Haldane, J. B. S. *Daedalus; or, Science and the Future*. New York: Dutton, 1925.

Hassett, Brenna. *Built on Bones: 15,000 Years of Urban Life and Death*. New York: Bloomsbury, 2017.

Havkin-Frenkel, Daphna, and Faith C. Belanger, editors. *Handbook of Vanilla Science and Technology*. Chichester, UK: Wiley-Blackwell, 2011.

Hewlett, Barry S., and Michael E. Lamb, editors. *Hunter-Gatherer Childhoods: Evolutionary, Developmental, and Cultural Perspectives*. New Brunswick, NJ: Routledge, 2009.

Hoffmann, Roald. *The Same and Not the Same*. New York: Columbia University Press, 1995.

Hofstadter, Richard. *Social Darwinism in American Thought*. Boston: Beacon Press, 1992.

Howard, Albert. *An Agricultural Testament*. London: Oxford University Press, 1940.

Hrdy, Sarah Blaffer. *Mother Nature: Maternal Instincts and How They Shape the Human Species*. New York: Ballantine Books, 2000.

Huesemann, Michael, and Joyce Huesemann, *Techno-Fix: Why Technology Won't Save Us or the Environment*. British Columbia: New Society, 2011.

Huxley, Thomas H. *Man's Place in Nature*. Mineola, NY: Dover, 2003.

Johnson, Nathanael. *All Natural*: *A Skeptic's Quest to Discover If the Natural Approach to Childbirth, Healing, and the Environment Really Keeps Us Healthier and Happier*. New York: Rodale, 2013.

Jütte, Robert. *Contraception: A History*. Translated by Vicky Russell. Malden, MA: Polity Press, 2008.

Kaye, Howard. *The Social Meaning of Modern Biology: From Social Darwinism to Sociobiology*. New York: Routledge, 2017.

Keeley, Lawrence H. *War before Civilization: The Myth of the Peaceful Savage*. New York: Oxford University Press, 1997.

Kermode, Frank. *The Sense of an Ending: Studies in the Theory of Fiction*. New York: Oxford, 1967.

Krech, Shepard. *The Ecological Indian: Myth and History*. New York: Norton, 2000.

Lakoff, George, and Mark Johnson, *Metaphors We Live By*. Chicago: University of Chicago Press, 2003.

Lamplugh, Rick. *In the Temple of the Wolves: A Winter's Immersion in Wild Yellowstone*. Self-published, CreateSpace, 2013.

Latz, Leo J. *The Rhythm of Sterility and Fertility in Women*. Chicago: Latz Foundation, 1936.

Laudan, Rachel. *Cuisine and Empire: Cooking in World History*. Berkeley: University of California Press, 2015.

Leclerc George-Louis, Count of Buffon. *A Natural History, General and Particular*. Translated by William Smellie. London: Thomas Kelly, 1856.

Leopold, Aldo. *A Sand County Almanac*. New York: Oxford University Press, 1966.

Levin, Yuval. *The Great Debate: Edmund Burke, Thomas Paine, and the Birth of Right and Left*. New York: Basic Books, 2014.

Lewis, C. S. *The Abolition of Man*. New York: Macmillan, 1953.

———, *Studies in Words*. Cambridge, UK: Cambridge University Press, 1960.

Lincoln, Bruce. *Theorizing Myth: Narrative, Ideology, and Scholarship*. Chicago: University of Chicago Press, 1999.

Lopez, Barry. *Of Men and Wolves*. 1978. Reprint, New York: Scribner Classics, 2004.

Lovejoy, Arthur. *The Great Chain of Being*. 1936. Reprint, Cambridge, MA: Harvard University Press, 1950.

Lutts, Ralph H. *The Nature Fakers: Wildlife, Science, and Sentiment*. Charlottesville: University Press of Virginia, 1990.

MacKinnon, J. B. *The Once and Future World: Nature as It Was, as It Is, as It Could Be*. Boston: Harcourt, 2013.

Mäki, Uskali, ed. *Fact and Fiction in Economics: Models, Realism, and Social Construction*. Cambridge, UK: Cambridge University Press, 2002.

Mann, Charles C. *1491: New Revelations of the Americas before Columbus*. New York: Vintage Books, 2011.

————. *The Wizard and the Prophet: Two Remarkable Scientists and Their Dueling Visions to Shape Tomorrow's World*. New York: Vintage Books, 2019.

Marris, Emma. *Rambunctious Garden: Saving Nature in a Post-Wild World*. New York: Bloomsbury, 2011.

Marx, Leo. *The Machine in the Garden: Technology and the Pastoral Ideal in America*. 1964. New York: Oxford University Press, 2000.

May, Elaine Tyler. *America and the Pill: A History of Promise, Peril, and Liberation*. New York: Basic Books, 2010.

Mill, John Stuart. *Three Essays on Religion*. London: Longmans, Green, Reader, and Dyer, 1874.

Mirowski, Philip. *More Heat than Light: Economics as Social Physics, Physics as Nature's Economics*. Cambridge, UK: Cambridge University Press, 1991.

Mirowski, Philip, ed. *Natural Images in Economic Thought: "Markets Read in Tooth and Claw."* Cambridge, UK: Cambridge University Press, 1994.

Morgan, Lewis H. *Ancient Society*. New York: Henry Holt, 1877.

Mumford, Lewis. *The Pentagon of Power*, vol. 2, *The Myth of the Machine*. New York: Harcourt, 1970.

Noonan, John T. *Contraception: A History of Its Treatment by Catholic Theologians and Canonists*. Enlarged edition. Cambridge, MA: Harvard University Press, 2012.

Odoux, Eric, and Michel Grisoni, editors. *Vanilla*. Boca Raton, FL: CRC Press, 2010.

Pascoe, Peggy. *What Comes Naturally: Miscegenation Law and the Making of Race in America*. New York: Oxford University Press, 2009.

Passmore, John. *Man's Responsibility for Nature*. London, Duckworth, 1974.

Pendergrast, Mark. *For God, Country and Coca-Cola: The Definitive History of the Great American Soft Drink and the Company That Makes It*. New York: Basic Books, 2013.

Peterson, Brenda. *Wolf Nation: The Life, Death, and Return of Wild American Wolves*. Boston: Da Capo Press, 2017.

Pielke, Roger, Jr. *The Edge: The War against Cheating and Corruption in the Cutthroat World of Elite Sports*. Berkeley, CA: Roaring Forties Press, 2016.

Pinker, Steven. *Enlightenment Now: The Case for Reason, Science, Humanism, and Progress*. New York: Penguin, 2019.

————. *The Blank Slate: The Modern Denial of Human Nature*. New York: Penguin, 2002.

Pollan, Michael. *In Defense of Food: An Eater's Manifesto*. New York: Penguin, 2009.

————. *The Omnivore's Dilemma: A Natural History of Four Meals*. New York: Penguin, 2006.

————. *Second Nature: A Gardener's Education*. New York: Grove Press, 1991.

Powers, Richard. *The Overstory*. New York: Norton, 2018.

Pritchard, James. *Preserving Yellowstone's Natural Conditions: Science and the Perception of Nature*. Lincoln: University of Nebraska Press, 1999.

Purdy, Jedediah. *After Nature: A Politics for the Anthropocene*. Cambridge, MA: Harvard University Press, 2015.

Pusey, James Reeve. *China and Charles Darwin*. Cambridge, MA: Harvard University Press, 1983.

Quammen, David. *Yellowstone: A Journey through America's Wild Heart*. Washington, DC: National Geographic, 2016.

Rain, Patricia. *Vanilla: The Cultural History of the World's Favorite Flavor and Fragrance*. New York: Penguin, 2004.

Redman, Charles L. *Human Impact on Ancient Environments*. Tucson: University of
    Arizona Press, 1999.

Reich, Jennifer A. *Calling the Shots: Why Parents Reject Vaccines*. New York: New York
    University Press, 2018.

Rocke, Michael. *Forbidden Friendships: Homosexuality and Male Culture in Renaissance Flor-
    ence*. New York: Oxford University Press, 1996.

Roosevelt, Theodore. *The Wilderness Hunter*. New York: Charles Scribner's Sons, 1923.

Ropeik, David. *How Risky Is It, Really? Why Our Fears Don't Always Match the Facts*. New
    York: McGraw-Hill Professional Publishing, 2010.

Ross-Bryant, Lynn. *Pilgrimage to the National Parks: Religion and Nature in the United
    States*. New York: Routledge, 2013.

Rush, Benjamin. *Medical Inquiries and Observations*, Vol. I. Philadelphia: Bennett and
    Walton, 1815.

Russell, Bertrand. *Icarus: Or the Future of Science*. New York: Dutton, 1926.

Ryan, Christopher, and Cacilda Jethá, *Sex at Dawn: The Prehistoric Origins of Modern
    Sexuality*. New York: Harper, 2010.

Salatin, Joel. *The Marvelous Pigness of Pigs*. New York: FaithWords, 2017.

Sapolsky, Robert M. *Behave: The Biology of Humans at Our Best and Worst*. New York:
    Penguin, 2017.

Schullery, Paul. *Past and Future Yellowstones: Finding Our Way in Wonderland*. Salt Lake
    City: University of Utah Press, 2015.

Scott, James C. *Against the Grain: A Deep History of the Earliest States*. New Haven, CT:
    Yale University Press, 2017.

———. *Seeing Like a State: How Certain Schemes to Improve the Human Condition Have
    Failed*. New Haven, CT: Yale University Press, 1998.

Seton, Ernest Thompson. *Wild Animals I Have Known*. New York: Charles Scribner's
    Sons, 1898.

Sideris, Lisa H. *Consecrating Science: Wonder, Knowledge, and the Natural World*. Oakland:
    University of California Press, 2017.

Smith, Adam. *An Inquiry into the Nature and Causes of the Wealth of Nations by Adam
    Smith, Edited with an Introduction, Notes, Marginal Summary and an Enlarged Index by
    Edwin Cannan*. 1776. Reprint, London: Methuen, 1904.

Smith, Douglas, and Gary Ferguson, *Decade of the Wolf: Returning the Wild to Yellowstone*.
    Guilford, CT: Lyons Press, 2012.

Stone, Geoffrey R. *Sex and the Constitution: Sex, Religion, and Law from America's Origins
    to the Twenty-First Century*. New York: Liverwright, 2017.

Tentler, Leslie Woodcock. *Catholics and Contraception: An American History*. Ithaca, NY:
    Cornell University Press, 2004.

Thomas, Chris D. *Inheritors of the Earth: How Nature Is Thriving in an Age of Extinction*.
    New York: Public Affairs, 2017.

Thomas, Keith. *Man and the Natural World: Changing Attitudes in England 1500–1800*.
    New York: Oxford University Press, 1996.

Tone, Andrea. *Devices and Desires: A History of Contraceptives in America*. New York: Hill
    and Wang, 2002.

Topinard, Paul. *L'Homme dans la nature*. Paris: Félix Alcan, 1891.

Unschuld, Paul U. *Medicine in China: A History of Ideas*. 2nd ed. Berkeley: University of
    California. Berkeley Press, 2010.

Van Gennep, Arnold. *The Rites of Passage*. Chicago: University of Chicago Press, 1960.
Vargas Llosa, Mario. *The Storyteller*. Translated by Helen Lane. New York: Penguin, 1990.
West, Geoffrey. *Scale: The Universal Laws of Life, Growth, and Death in Organisms, Cities, and Companies*. New York: Penguin, 2017.
Wexler, Philip. *History of Toxicology and Environmental Health: Toxicology in Antiquity*, Volume II. London: Elsevier, 2015.
Whorton, James C. *Nature Cures: The History of Alternative Medicine in America*. New York: Oxford University Press, 2002.
Wicker, Brian. *The Story-Shaped World: Fiction and Metaphysics; Some Variations on a Theme*. Notre Dame, IN: University of Notre Dame Press, 1975.
Wiener, Norbert. *The Human Use of Human Beings: Cybernetics and Society*. London: Eyre and Spottiswoode, 1954.
Williams, Joy. *Ill Nature: Rants and Reflections on Humanity and Other Animals*. New York: Vintage Books, 2002.
Williams, Raymond. *The Country and the City*. New York: Oxford University Press, 1973.
———. *Culture and Society: 1780–1950*. New York: Columbia University Press, 1960.
———. *Keywords: A Vocabulary of Culture and Society*. London: Flamingo, 1983.
Wills, Garry. *Papal Sin: Structures of Deceit*. New York: Doubleday, 2002.
Wilson, Bee. *Swindled: The Dark History of Food Fraud, from Poisoned Candy to Counterfeit Coffee*. Princeton, NJ: Princeton University Press, 2008.
Woloson, Wendy. *Refined Tastes: Sugar, Confectionery, and Consumers in Nineteenth-Century America*. Baltimore: Johns Hopkins University Press, 2002.

Index

abortion, 26, 151, 171, 179, 180
Abreu, Tifanny, 204
abstinence, 172, 173, 174, 230n2. *See also*
    natural family planning (NFP); sex
    and sexuality
Aché, 25
Act to Protect the Birds and Animals
    (1894), 93
acupuncture, 108, 114
adultery, 171, 186
Affluence without Abundance (Suzman),
    62–63
Afghanistan, 24
Against the Grain (Scott), 60
agricultural labor, 37, 50–51
agricultural research, 35–37, 43, 51, 54–55
An Agricultural Testament (Howard), 208
agriculture, 1–2, 55–56, 207–8, 211
Agta, 25
AIDS epidemic, 6
air pollution, 221n24. *See also* toxic chemi-
    cals and health
Aka, 26
Alaffia, 134–35
Albius, Edmond, 42
Albright, Horace, 90
alcoholic beverages, 45–46, 47, 58
algae farming, 36
altruism and branding, 134–35
ambiguities *vs.* certainty, 205–13
America (publication), 143
American Journal of Bioethics (publication),
    203
American Wolf (Blakeslee), 96
Amish, 61
Ammous, Saifedean, 154, 155
Anabolic Steroid Act (2004), 193

anabolic steroids, 191–93, 194
animal agriculture. *See* farming
animals (nonhuman): birthing experiences
    of, 23–25; in captivity, 24, 29, 58;
    cloning of, 11, 206, 232n36; federal
    protections for, 93, 98–99, 100–101;
    same-sex relations in, 187–88. See also
    names of specific species
anthropology, as discipline, 65
anthropomorphism, 91–92, 94, 96–97
Apicius (cookbook), 46
"appeal to nature" fallacy, 8, 216n14
Aristotle, 156–57, 169, 230n42
artificial and synthetic flavoring, 39, 43,
    44–45
As Nature Leads (Rogers), 186
aspartame, 45
astrology, 15
athletes: equipment/tools of, 195–96; as
    natural *vs.* unnatural, 190–95; sexism
    and sex segregation of, 6, 199–205;
    transgender athletes, 203–5
Atkin, Emily, 145
Augustine (saint), 172, 179
ayurveda, 108, 125, 133
Aztec myths, 58

Baartman, Sara "Saartjie," 21
baboons, 25
*The Backward Peoples and Our Relations to
    Them* (Johnston), 21
Baker, Shawn, 153–54
Bakker, Jim, 151
balance, 37, 75–76, 115, 120
Bau, 65
Bäverhojt, 53
bears, 85, 88, 89–91, 93

Becerra, Xavier, 144
Becker, Carl, 227n4
Berenstein, Nadia, 48
Berkeley, George, 161
bestiality, 179, 184, 185, 231n22
beverages, 45–46
Bhai, Ferdous, 154–55
Biblical stories, 12, 15
Bigelow, Jacob, 117
biophilia, 82–83
birth control methods, 171–83, 230n2
birth control pill, 172, 174, 181
birth stories, 15–16. *See also* creation myths
bison, 99, 100–103
Biss, Eula, 30
bitcoin, 154–56
The Bitcoin Standard (Ammous), 154, 155
black bears, 90
Blackfoot, 102
Blackstone, William, 179, 231n22
black women and medical care, 21. *See also* women
Blakeslee, Nate, 96, 97
Blondiau, Eloise, 143
blood-doping, 193–94
Blurton-Jones, Nicholas, 76
body as model for society, 157–62
bodybuilding, 190–93
Bond, Timothy, 64
Book of Esther, 139
Boone and Crockett, 105
Born in the Wild (video), 19
Bororo, 67
Bourdain, Anthony, 74
Bowlby, John, 30–31
boxing, 200
Bradley, Robert, 23–24
Bradley Method of Natural Childbirth, 17, 18
Brazil, 32
breastfeeding, 29–30
British Journal of Sports Medicine (publication), 193
British Medical Journal, 129
Brooks, Carol Flora, 180–81
Brown, Louise, 12
Buckley, William F., Jr., 163
Buffalo Field Campaign, 103, 104
Buffon, Comte de, 21
Lafayette Bunnell, 86
Burke, Edmund, 159–60, 164, 229n16

Burlamaqui, Jean Jacques, 149
Burroughs, John, 91–92
Bushmen of the Kalahari, 62–63, 64

$^{13}C/^{12}C$ (isotope), 49–50
cacao production, 37, 45
caesareans, 16, 17, 32. *See also* childbirth
calendrical nomenclature, 80
Cameron, Catherine, 62, 65
Campbell, Joseph, 112
camping, 87–90, 92
Canada, 107–8, 128–29
cancer, 107–8, 113, 120, 123, 130
Candler, Asa, 45
candles, 133, 138–39
capitalism, 6, 129, 167
captive animals, 24, 29, 58. *See also* animals (nonhuman)
captivity, humans in, 62
Carlsen, Magnus, 196
Carnegie, Andrew, 163
carnivory, 153–55
caste system, 137–38
castoreum, 53
Catholicism: on birth control and sex, 171–74, 176–78, 181, 183–84; on interracial marriage, 186; miracle stories in, 127; moral safety in, 148
C. B. Woodworth Sons Company, 44
Cederström, Carl, 138
celibacy, 175. *See also* sex and sexuality
certainty *vs.* ambiguity, 205–13
Chace, E. M., 35
Chand, Dutee, 202–3
Chase, Alvin W., 45
chess, 196–97
chickens, 2, 5, 58, 211
childbirth: animals experiences of, 23–25; caesareans in, 16, 17, 32; creation stories and, 15; in developing nations, 19, 21–22; historical accounts of, 16, 20–22; among hunter-gatherer groups, 25–26, 33; maternal mortality in, 22, 24, 25; modern obstetric advancements in, 22, 26; myths and, 33–34; neonatal mortality, 22, 24; prehistoric evidence of, 26–27; principles on natural methods, 16–20, 27–28; use of scientific metrics and, 30, 31–32, 34. *See also* infant mortality; nature and natural; women

child-rearing, 69–71, 77–78
children: agricultural labor by, 37, 50–51;
    breastfeeding of, 29–30; infant mortal-
    ity, 24, 25, 176–77; neonatal mortal-
    ity, 22, 24; raising of, 69–71, 77–78;
    schools for, 72, 73
Chile, 26–27
Chinese traditions: on Darwin, 165–66;
    in medicine, 81, 110, 112, 114, 115,
    125; on *ziran* and natural living, 3–4,
    66–67, 221n24
Chopra, Deepak, 113, 131–32, 141
Chopra Center, 140
Christian Family Movement, 183
Christian Science, 177
Clement, Brian, 107, 108–9, 125, 128
cloning, 11, 206, 232n36
Coatlicue, 58
Coca-Cola, 45, 58
cocaine, 45, 219n13
coca leaves, 45, 219n13
Cohen, I. Bernard, 161
coitus interruptus, 172, 174–75, 183. *See also*
    birth control methods
Coleman, Ronnie, 191
Colistra, Joe, 144
colonialist privilege, 21–22
Columbus, Christopher, 69
communism, 6, 166, 167, 168
Comstock Laws (US), 180
condoms, 172, 173, 174–75, 176. *See also*
    birth control methods; contraception
Confucius, 221n24
Conger, Omar, 101
consecrated consumption, 135–46
Consumer Reports, 52
contraception, 171–83, 230n2
Cooks Illustrated, 49
Copernicus, 81
corn, cultivation of, 150, 152
cortisone, 193–94
cosmetics, 133–34, 138–39
Coubertin, Pierre de, 199
Coughlin, Charles, 178
coumarin, 48
Cox, Ignatius, 178
creation myths, 15, 33–34, 38, 124, 137.
    See also birth stories
"crimes against nature," 179–81
CRISPR-Cas9 technology, 206
cryptocurrency, 154–56

crystals, 145
Cuisine and Empire (Laudan), 138
cultural tourism, 71
Cupid's Yokes (Heywood), 174–75, 180
Cusma, Elisa, 201
Cuvier, Georges, 21
Cuyahoga River, 122

Dalkon Shield, 175
Dante, 99
Danwatch, 50–51
Darwin, Charles, 4–5, 21, 163
Darwinism, 4–5, 162–68
Davis, Hugh J., 175
death, 130
Decade of the Wolf (Smith and Ferguson),
    96
Declaration of Independence, 149
deforestation, 51, 221n24
defrutum, 45, 46, 47
Delos, 131–32, 134, 145
depression, 65–66, 74–75, 76, 128
Derkatch, Colleen, 126
The Descent of Man (Darwin), 163
descriptive laws, 148
detoxification, 139–40
diagnostic rituals, 120–21, 122–23
Diamond, Jared, 70
DiCaprio, Leonardo, 134
Diet Coke, 58
diets: carnivory, 153–54; Paleo, 26, 154,
    155; veganism, 108, 154; vegetarian-
    ism, 56, 114. *See also* food
Dillard, Annie, 27
Discovery of the Yosemite (Bunnell), 86
disgust, 183–84, 188
Dogon, 176
dogs, 24
Dolly (sheep), 11
Don Alberto, 72, 73, 74
Douglas, Mary, 141
Draper, Patricia, 65
Drummond, Henry, 147–48, 149–51
Dryden, John, 59, 66
Dunsworth, Holly, 29–30

The Ecological Indian (Krech), 61
economics and nature: Aristotle on,
    156–57, 230n42; Burke on, 159–60;
    cryptocurrency, 154–55, 228n5; evolu-
    tionary biology on, 162–68; principles

of interest, 169; Rousseau on, 160–61; Smith on, 158–60, 169. *See also* income inequality; nature and natural
ecosystem, as concept, 99
Edward, Gethin, 124
Edward, Prince of Wales, 116
Egyptian traditions, 67
Ehrenreich, Barbara, 123, 130
Eliade, Mircea, 67
elimination communication, 70, 78
elk, 104
Ellingson, Ter, 68, 75
Emerson, Ralph Waldo, 162
Émile (Rousseau), 56
Ending Medical Reversals (Prasad), 129
Engels, Friedrich, 166
Environmental Protective Agency (EPA), 122
"environment of evolutionary adaptedness," 31–32, 218n34
Epicurious (blog), 49
epilepsy, 113
Equal Rights Initiative, 185
Errington, Paul, 87
etymology, 16
eugenics, 6, 162–63, 229n21
Euripides, 22
Evelyn, John, 221n24
evolutionary biology, 4–5, 27, 162–68
Evolva, 52

fake meat, 1, 206
farming, 1–2, 55–56, 207, 211
FDA (US Food and Drug Administration), 5, 51, 53
feminism, 31
Fergren, Gary, 112
fermentation, 51–52
fertility awareness based methods (FABMs), 230n2. *See also* natural family planning (NFP)
Finn, Robin, 132
Finnis, John, 227n4
Fire Lame Deer, John, 100, 104
First Nations. *See* indigenous groups
Fischer, Simona, 145
fish, 36, 58, 221n24
flamingo recipe, 46
food: artificial and synthetic flavoring of, 39, 43, 44–51; farm production model, 1–2, 55–56; FDA regulations on, 44,

51; lab-grown meat, 1, 206; natural food as regulated term, 5, 53–54; organic, 207–8; rituals of, 183–84; Roman vomitoriums and, 177, 178, 184; social class and, 138; (un)natural principles on, 56–58. *See also* diets; nature and natural; vanilla
Foote, Mary Hallock, 172
Forrest, Darci, 39
The Four Agreements (Ruiz), 113
Francis (pope), 171–72, 177
Franke, Mary Ann, 104–5
Frankenfood, 11
Frankenstein (Shelley), 11–12, 14
"freak of nature," as phrase, 191–92
Friedman, Milton, 163
Friends of the Earth, 52
Fry, Douglas, 61
Fumifugium (Evelyn), 221n24

Galton, Francis, 162–63
Gaskin, Ina May, 18, 20, 24
Gautier, Émile, 164–65
gender. *See* women
gene-editing technology, 206
Genesis, 12, 15
genetically modified organisms. *See* GMO technology
Georgia, 186
gibbons, 24
Gladwell, Malcolm, 174, 176
GMO technology, 2–3, 11, 52, 207, 210–11
Golf Digest, 193
Goop, 133–34, 139, 140–41, 144, 145
Gordon, Linda, 175
gorillas, 24
Gould, Stephen Jay, 148, 150, 151, 152
Grabowski, Alena, 198, 204
Graham, Sylvester, 56, 118, 225n16
grandmother hypothesis, 78
Grant, Ulysses S., 101
gray whales, 103–4
Great Britain, 158
Greek traditions: in medicine, 110, 111, 112–13; myths of, 12, 34, 67; political and economic systems, 4, 156
Greene, Maurice, 193
Gregory, Alex, 60
Gregory XI (pope), 169
Grigg-Spall, Holly, 175–76, 182
grizzly bears, 85, 88, 93

Grizzly Man (Herzog), 88
growth hormones, 191, 192–93
guaiacol, 50
Guardian (publication), 51
Guatemalan midwifery, 20
Gunter, Jennifer, 139

Häagen-Dazs, 52
Hadza, 76
Haeckel, Ernst, 165
Hall, Lenwood, 192
Hamblin, James, 144
Hammond, Jay, 94
Hansen, Michael, 52
happiness studies, 63–64
Hari, Vani, 53–54
Harper, Joanna, 204, 205
Hatch, Orrin, 188
Haudenosaunee, 107–8, 124–25
Hawkes, Kristen, 78
Healing Cancer (film), 109, 120
The Healing Hand (Weltmer), 117
Healing Our Children (Nagel), 20
health. *See* illness; medicine; natural
    medicine
Heath, Phil, 191
Henderson, L. J., 117
hermit crabs, 147–48
Hershey's Kisses, 39
Herzog, Werner, 88–89
Hewlett, Barry, 26
Heywood, Ezra, 174–75, 180
Hicks, Thomas, 195
high-fructose corn syrup, 39, 45
Hill, Lisa, 159
Hinduism, 67, 137–38
Hippocrates, 110, 112–14
Hippocrates Health Institute, 107–9, 114,
    120, 122, 125, 126–28
Hiwi, 25
Hobbes, Thomas, 59, 66
Hoeve Biesland, 55
Hoffman, Roald, 49, 56–57
Hofstadter, Richard, 163
Holland, 35–37
Holmes, Oliver Wendell, Sr., 117
Holt, James, 102
Holy Sex (Popcak), 182–83
homosexuality, 7, 171, 179, 187–88,
    232n52
Hopi, 15, 33

hormone treatments, 204
Horn, Trent, 182
horses, 24
housing, 131–32, 140, 141, 144
Howard, Alfred, 208
Hrdy, Sarah Blaffer, 31–32
Huacaria, Peru, 69, 70–75
Huainanzi, 67
Humanae vitae (Paul VI), 171, 173–74
human body, 157–62
humility, 212
hunter-gatherer groups, 25–26, 33, 62–65,
    75–78. *See also* indigenous groups
hunting, 90–91, 93, 98, 102–5
Huxley, Thomas, 21
hyenas, 24
Hyman, Mark, 54

IAAF (International Association of Athlet-
    ics Federations), 195, 197–203
ideological monoculture *vs.* polyculture,
    208–9
Ik, 75
Iliad (Homer), 111
illness, 115–18, 123, 136–37, 140. *See also*
    cancer; natural medicine
The Illness Narratives (Kleinman), 115
Impossible Burger, 1
Ina May's Guide to Childbirth (Gaskin), 18,
    21
income inequality, 66, 136–37, 140–46. *See
    also* economics and nature
In Defense of Food (Pollan), 40
India: sports in, 202; traditions of, 33, 67,
    108, 137–38
indigenous groups: birth stories of, 15,
    20, 124; childbirth and, 19, 21–22;
    dehumanization of, 67–69; environ-
    mental practices of, 61–62; hunting
    ritual of, 98, 100; medicine and health
    of, 107–8, 124–25, 128–29; in Peru,
    69, 70–73; wilderness protection
    and, 99–101. *See also* hunter-gatherer
    groups
Indonesia, 43
infant mortality, 24, 25, 176–77
Inferno (Dante), 169
Information for Everybody (Chase), 44–45
In Goop Health symposium, 139
interdependence, 99
interest (economic), 169

International Association of Athletics Federations (IAAF), 195, 197–203
International Federation of Bodybuilding and Fitness, 191
International Olympic Committee (IOC), 200–201, 203
interracial marriage and sex, 6, 185–87
InterTribal Bison Cooperative, 105
In The Temple of the Wolves (Lamplugh), 97
Inupiat Eskimos, 95
invasive plants, 209–10
Iroquois. *See* Haudenosaunee
Islamic law, 169
IUD (intrauterine devices), 172, 175
Izquierdo, Carolina, 70–71

Jefferson, Thomas, 149, 150
Jelinger, Christopher, 169
J.J., 107–8, 123–24
Jocks, Christopher, 125
"John Rock's Mistake" (Gladwell), 174, 176
Johnson, Candace, 21
Johnson, Mark, 12
Johnson, Nathanael, 24
Johnston, Henry, 21
Judaism, 80
Juma, Calestous, 7, 9
Just, Adolf, 119

Karkazis, Katrina, 202
Kass, Leon, 232n36
Kaye, Howard, 167
Keeley, Lawrence, 62
Kellogg, John Harvey, 118, 119
Kepler, Johannes, 81
Kerr, Roslyn, 200
"The Killing Game" (Williams), 103–4
King, Helen, 114
King, Martin Luther, Jr., 68
Kipchoge, Eliud, 196
Kipling, Rudyard, 68
Kleinman, Authur, 115
Knaus, Hermann, 172, 173
Kolbert, Elizabeth, 70
Konner, Melvin, 26, 32–33
Krauthammer, Charles, 11
Krech, Shepard, 61
Kropotkin, Peter, 165, 167
!Kung San, 63, 65

lab-grown meats, 1, 206
Lady Gaga, 187
Lakoff, George, 12
Lakota, 100
Lamarck, Jean-Baptiste, 166
Lamarckian evolutionary theory, 166–67
Lamaze, Fernand, 16
Lamaze International, 16, 17, 18
Lambert Snyder Vibrator, 119
Lamplugh, Rick, 97, 103
Lancet (publication), 44
Lang, Kevin, 64
Lange, Lisa, 11
Langston, Nancy, 209–10
Latz, Leo J., 172, 177–79
Laudan, Rachel, 63, 65, 138
Lead and Lead Poisoning in Antiquity (Nriagu), 46
lead exposure, 46
Leave No Trace principles, 98–99
Le Darwinisme Social (Gauthier), 164–65
Lee, Richard B., 63
Lengfeld, A. L., 48
Levitt, Arthur, 155, 228n5
Lewis, Paul, 11
lifespan, 60, 62, 76
Lincoln, Bruce, 12
lions, 24
The Living Temple (Kellogg), 118–19
"Lobo, the King of Currumpaw" (Seton), 95–96
Long, William J., 91
Lopez, Barry, 94
López-Alt, J. Kenji, 49
Lorz, Fred, 194–95
Luther, Martin, 142, 146
Luther Standing Bear (chief), 99
Lyell, Charles, 167
lynchings, 184–85

Machar, Agnes Maule, 116
Madagascar, 43, 50–51
magical medicine, history of, 111–12
Makah, 103
makeup, 133–34, 138–39
The Mama Natural Week-by-Week Guide to Pregnancy and Childbirth (Howland), 18
Mann, Charles C., 208
Mäntyranta, Eero Antero, 192

Mao Tse-tung, 166
marathon runners, 194–96. *See also* running
"March of Progress" parody, 59
Maritain, Jacques, 181
marriage rituals, 79–80
Martínez-Patiño, Maria José, 201–2
Marx, Karl, 166
Massachusetts, 180
Massachusetts Medical Society, 117
maternal mortality, 22, 24, 25. *See also* childbirth
Matsigenka, 69, 70–75, 77
Mayans, 41–42
Mbeki, Thabo, 6
McDonald's, 39–40
McIntyre, Rick, 97–98
meat substitutes, 1, 206
Medea (Euripides), 22
medical materialism, 140–41
medicine: Chinese traditions on, 81, 110, 112, 114, 115, 125; Greek traditions on, 110, 111, 112–13; modern medical science, 22, 26, 120–25. *See also* childbirth; natural medicine
Mencius (Confucius), 221n24
Mencken, H. L., 171
menstrual cycle, 172, 176. *See also* natural family planning (NFP)
Mexico, 38
Meyer, Joseph E., 118, 119
Michtom, Morris and Rose, 90, 91
Mill, John Stuart, 8
Mills, Ashea, 88
Milne, A. A., 91
miscegenation, 185–86
modeling, 228n8
modernity: health and, 118, 121–22; hunter-gatherer groups and, 64–66; Matsigenka and, 70–75; parenting and, 69–71; wilderness and, 105–6
modern medical science, 22, 26, 120–25. *See also* medicine; natural medicine
Mohawk, 125
money laundering, 51
monkeys, 25
monoculture *vs.* polyculture, 208–9
monogamy, 175, 188
monopolies, 160, 169
Montana Department of Fish, Wildlife and Parks, 102, 103

Montreal Daily Star (publication), 199
moral law *vs.* natural law, 147–52, 227n4
moral principles of nature, 147–48
Mr. Olympia competition, 191–92
murder, 4, 171, 179, 184–85
Muse Residences, 131–32
mutual aid, 165, 167–68
Mutual Aid (Kropotkin), 165
The Myth of the Noble Savage (Ellingson), 68, 75
myths, 11–14, 33–34, 112. *See also* religion

Nagel, Ramiel, 20
names and rituals, 79–80
The Nasadiya Sukta (Rigveda), 212–13
National Aboriginal Health Organization (NAHO), 124
National Academy of Sciences, 148
nationalism, 6
national parks: camping in, 87–90; federal animal protections in, 93, 98–99, 100–101; rituals within, 92–93; wilderness, defined, 98–99. *See also* Yellowstone National Park
Native Americans. *See* indigenous groups
natural childbirth principles, 16–20, 27–28. *See also* childbirth
natural family planning (NFP), 172, 173, 176, 181, 183, 230n2. *See also* birth control methods; sex and sexuality
Natural Law in the Spiritual World (Drummond), 147–48, 149–50
natural law *vs.* moral law, 147–52, 227n4
natural medicine: history of, 110–15; modern medical science on, 120–25; qualities and principles of, 123–30; religious language and, 115–18, 127; as term, 112; using food in, 107–10. *See also* medicine; nature and natural
natural personal care products, 133–36
natural selection. *See* Darwinism
nature and natural, 1–9; Aristotle on, 230n42; balance of, 115; Catholics on birth control as, 171–74; consecrated consumption and, 135–46; as cultural construct, 6, 22–23, 121, 189; definition of, 12–13, 38, 51, 143, 191; etymology of, 16; fallacy of, 7–8, 216n14; "freak of nature," as phrase, 191–92; hunting and, 90–91, 93, 94,

98, 102–5; inaccurate (fake) depictions of, 89–92; modernity and rituals in, 82–83, 105–6; moral principles of, 147–48; myths on, 11–14; pilgrimages to, 86–87; Pollan on, 219n4; as religious term, 4; scientific metrics *vs.*, 30, 31–32, 34, 120–25. *See also* animals (nonhuman); childbirth; economics and nature; food; sex and sexuality; vanilla
nature-deficit disorder, 82
Nature Delineated (Pluche), 150, 227n7
"Nature Fakers" (Roosevelt), 92
Nature Knows No Color-Line (Rogers), 186
Nau, Louis, 178
Nayaka, 26
neonatal mortality, 22, 24
Netherlands, 35–37
New Coke, 45
New Republic (publication), 145
Newsweek (publication), 127
Newton, Isaac, 161
New Yorker (publication), 49, 60, 69
New York Evening Post (publication), 199
New York Times (publication), 1, 11, 92, 132, 204
Nez Perce, 102, 103
Nigeria, 64, 75
Nike, 196
Nineteen Eighty-Four (Orwell), 13
noble savage myth, 32, 59, 65, 66–69. *See also* indigenous groups
nomenclature, 79–80
nonhuman animals. *See* animals (nonhuman)
Nonnatus, Raymond, 33
non-overlapping magisteria, 148
non-toxic branding, 133–34
Northern Cheyenne, 99–100
"Notes on the Original Affluent Society" (Sahlins), 63
Nriagu, Jerome O., 46
nuclear energy, 210
nuclear weapons, 64, 66
Nutella, 50

Oatman, McCoy, 103
Obama, Barack, 60–61
obesity, 31, 66
obstetric advancements, 22, 26. *See also* childbirth

The Official Lamaze Guide (Lothian and DeVries), 18, 19
Of Wolves and Men (Lopez), 94
Ogino Kyusaku, 172, 173
Olympic Games, 192, 194, 197, 199–200, 204
On Ancient Medicine (Hippocrates), 114
O'Neal, Shaquille, 192
On the Origin of Speces (Darwin), 21, 164, 166
On the Sacred Disease (Hippocrates), 113
Opp, James, 116
orangutans, 24
Organization of Competition Bodies (OCB), 190–91
origin stories. *See* creation myths
Orwell, George, 139
Osteen, Joel, 142
Our World in Data (publication), 64
overfishing, 221n24

Paine, Thomas, 164
Paleo diet, 26, 154, 155
Paltrow, Gwyneth, 133–34. *See also* Goop
Papua New Guinea, 19
Paraguay, 25
parenting, 69–71, 77–78
Parker, Theodore, 68
Pascoe, Peggy, 185–87
patient, as term, 123
Paul VI (pope), 171, 173–74, 177, 181
Pawnee, 99–100
peaceful societies, 61
Pemberton, John, 45
Pennsylvania, 185
penny candies, 44
People (publication), 70
People for the Ethical Treatment of Animals (PETA), 11
perfectionist theologies, 58, 59. *See also* nature and natural
performance-enhancement of athletes, 190–95, 204
perfumes, 133, 138–39
personal care products, 133–36
Peru, 69–75, 77
pesticides, 58, 207
Peterson, Brenda, 95
pew rental, 142–43
Phelps, Michael, 192
Philippines, 25, 171

Philistines, 111–12
physics, 81, 161–62
Pielke, Roger, Jr., 203
pigs, 2, 11
pilgrimages, 86–87
Pilgrimage to the National Parks (Ross-Bryant), 86–87
Pilgrim at Tinker Creek (Dillard), 27
Pill. *See* birth control pill
Pinker, Steven, 60, 66
Pistorius, Oscar, 197
Pius XI (pope), 173
plague, 112
Planned Parenthood, 187
plants, 128, 150, 152, 209–10. *See also* food; vanilla
Plato, 13–14, 156, 157
Pliny the Elder, 46, 47
Pluche, Noel Antoine, 150, 227n7
Polis, Chelsea, 230n2
political polarization, 151
political system as body, 157–58, 160–62
Politics (Aristotle), 156
Pollan, Michael, 2, 3, 40, 43, 219n4
pollution, 209, 221n24. *See also* toxic chemicals and health
polyculture *vs.* monoculture, 208–9
Polyface Farm, 1–2, 211
Pontifical Commission on Birth Control, 183
Popcak, Gregory, 182–83
Popper, Ilona, 85, 96–97
pornography, 180
poverty. *See* income inequality
PPL Therapeutics, 11
Prasad, Vinay, 120–21, 122, 129
prayer, 116
prehistoric evidence of childbirth, 27
Premium Body Care line, 134–35
prescriptive laws, 148
primary polycythemia, 192
primates, 24–25
Pritchard, James, 90
prosthetic *vs.* biological limbs of athletes, 197–98
Pure Food and Drug Act (1906), 44
purification and purity, 137–39, 140, 145–46. *See also* consecrated consumption
Purity and Danger (Douglas), 140
Pusey, James, 165–66
Push Back (Tuteur), 22

racism, 21, 184–85. naturalistic racism, 151. *See also* interracial marriage and sex; slavery
Raquel, Thomas, 48
raw water branding, 140, 145
"Real and Sham Natural History" (Burroughs), 91–92
Rehm, Markus, 197
Reich, Jennifer, 6, 121, 126, 130
reincarnation, 25
Reiner, Joseph, 173
religion: commodification of, 141–43; economic priniciples of, 168–69; medicine and, 112–13, 115–18, 127; nature and, 4–6, 8; rituals of, 80–81, 140; *vs.* science, 148–49; on sexual practices, 169–70. *See also* myths
reproductive fitness theories, 164, 229n21. *See also* Darwinism
Republic (Plato), 13–14, 156
Return to Nature! (Just), 119
rhythm method. *See* natural family planning (NFP)
The Rhythm of Sterility and Fertility in Women (Latz), 172–73, 177-78
Rights of Man (Paine), 164
Rigveda, 137, 211, 212–13
rituals, 79–82, 86–87, 115, 140
Roberts, B. T., 142–43
Roberts, David, 144
Rock, John, 174, 176
Rockefeller, John D., Jr., 153
Roddenbery, Seaborn, 185
Rogers, J. A., 186, 187
Roman Catholicism, 148
Roman Empire, 45–46, 177, 178
Roosevelt, Alice, 195
Roosevelt, Theodore, 90–91, 92, 102, 105
Rosenberg, Karen, 27, 32
Roser, Max, 64
Ross-Bryant, Lynn 86
Rousseau, Jean-Jacques, 56, 75, 160–61
Ruiz, Don Miguel, 113
Ruiz, Rosie, 195
Rule 144.3(d) (IAAF), 195, 197
running, 194–96, 197–98, 199
Rush, Benjamin, 20

Sahlins, Marshall, 63, 65
Salatin, Joel, 1–3, 5, 211

same-sex relationships, 7, 187–88, 232n52.
     See also homosexuality
Sampson, Jeffrey, 102
sapa, 45, 46, 47
Sapolsky, Robert, 62, 221n22
śarīra, 127
Sarma, Himanta Biswa, 81
Sasse, Ben, 151
Satya Yuga, 67
Sault, Makayla, 107–9, 125–26
To Save the Wild Bison (Franke), 104–5
Schmidt, Oscar, 168
Schnabel, Charles, 108
Schullery, Paul, 86
Schuyler, George, 186, 187
science *vs.* religion, 148–49
Scott, James, 60, 69, 74
seagulls, 187
secular purity rituals, 140–41
Semenya, Caster, 201–3
Sensenbrenner, Jim, 193, 194
Serious Eats, 49
Seton, Ernest Thompson, 91, 95–96
The Seven Spiritual Laws of Success (Chopra),
     141
Sex and Race (Rogers) 186
sex and sexuality: abstinence, 172, 173, 174,
     230n2; adultery, 171, 186; bestiality,
     179, 184, 185, 231n22; birth control
     methods, 171–83, 230n2; celibacy,
     175; "crimes against nature" on,
     179–81; homosexuality, 7, 171, 179,
     187–88, 232n52; interracial marriage
     and, 6, 185–87; monogamy, 175,
     188; moral disgust of, 181–84, 188;
     pornography, 180; sodomy, 169, 170,
     179–80
Sex at Dawn (Ryan and Jethá), 188
sexism and sex segregation in sports, 6,
     199–205
Shakespeare, William, 4, 15, 94
sharia law, 169
sheep, 11
Sheepeaters, 99
Shelley, Mary, 11–12, 14
Shelley, Percy, 56
shoulder dystocia, 20
Sierra Leone, 24
Silverstone, Alicia, 70, 78
Sitting Bull, 100
ski jumping, 199

slavery, 62, 156–57
Smith, Adam, 6, 158–59, 160, 161
Smith, Doug, 87, 94, 95, 102
Smith, Hannah, 28–29
soaps, 134–35
Social Darwinism, 162–68
Social Darwinism in American Thought
     (Hofstadter), 163–64
socialism, 166
A Society of Wolves (McIntyre), 98
Socrates, 156
sodomy, 169, 170, 179–80
South Africa, 6
Spade, Kate, 74
Spang-Willis, Francine, 99–100
Spencer, Herbert, 167
Spencer, Richard, 151
Spicer, André, 138
spiritual healing, 116
spontaneous abortions, 26. *See also* abortion
sports. *See* athletes
Stalin, Joseph, 153, 166
Starr, Oliver, 96
state of nature: ancient records on, 66–67;
     hunter-gatherer groups and, 62–65;
     modern tribes and, 69–72; Obama on,
     60–61; parodies on, 58–59. *See also*
     nature and natural
Statham, Travis, 154
Steinbeck, John, 57
The Stepan Company, 219n13
Steptoe, Patrick, 11–12
steroids, 191–93, 194
stillbirths, 25, 26–27
The Storyteller (Vargas Llosa), 77
Strassman, Beverly, 176
suffering, 65–67
suicide, 74, 76, 128
Summers, Larry, 162
Sun Yat-sen, 166
supernatural medicine, history of, 111–12.
     See also natural medicine
supplements (nutritional), 110, 121, 125,
     190–91
Sustaining Lake Superior (Langston), 209
Suzman, James, 62, 64
Swanson, Kevin, 81
Sweetening the Pill (Grigg-Spall), 175–76
swimming, 200
Swindled (Wilson), 47
SynchroDestiny, 141–42

taboos, 82
Talib, Nassim, 155
TallBear, Kimberly, 61
Tanzania, 25
tattoos, 178
teddy bear, 90–91
Teniztli, 38
Tercek, Michael, 88
test tube baby, 11–12
Thomas, Elizabeth, 61
Thoreau, Henry David, 7, 98
Thrash, Marjorie, 190, 192
Thurber, Simone, 19
Time (publication), 11
tobacco, 47
Tohono O'odham, 111
tonka bean, 48
Tourmaline Spring of Maine, 140, 145
Toxic Bodies (Langston), 209
toxic chemicals and health, 133–36, 209.
    See also pollution
transgender athletes, 202–5. *See also*
    athletes
Treadwell, Timothy, 88–89
trust, 129
Tshabalala-Msimang, Manto, 6
Tucker, Ross, 200
Tuteur, Amy, 22
Tyndall, John, 116–17
Tzacopontziza, 38

Uganda, 75
Ulrich, Roger, 128
umbilical cord-cutting, 33
unnatural *vs.* natural, overview, 1–9. *See also*
    nature and natural
US Food and Drug Administration (FDA),
    5, 51, 53
US Securities and Exchange Commission
    (SEC), 155
usury, 169
utopianism, 76–77

vaccinations, 6, 121, 126
Van der Meer, Theo, 179
vanilla: creation myths of, 38; cultivation
    and processing of, 35–37, 42–43,
    50–51, 54–55; history of, 41–42; in
    McDonald's and Hershey's products,
    39–40; synthetic forms of, 39, 43, 47,
    48–53. *See also* food

vanillin (synthetic vanilla), 39, 43, 47,
    48–53
vanillism (illness), 47
Van Noort, Filip, 35–37, 43, 51, 54–55
Vargas Llosa, Mario, 77
veganism, 108, 154
vegetarianism, 56, 114
Vice (publication), 154
Victoria (queen), 116
"A Vindication of Natural Diet" (Shelley),
    56
A Vindication of Natural Society (Burke),
    229n16
Virgil, 169
Virginia, 186
vis medicatrix naturae, 113, 117. *See also*
    natural medicine
Volney, Constantin François de Chasse-
    bœuf, comte de, 150
voluntary abstinence, 172, 173, 174, 230n2
vomitoriums, 177, 178, 184
Vox (publication), 144

Walden (Thoreau), 7
war, 61–62, 64–65
War before Civilization (Keeley), 62
Washington Post, 91
water, 140, 145–46
Watts, Jonathan, 51
wealth. *See* income inequality
The Wealth of Nations (Smith), 158–59
weight lifting, 200
wellness, 138–39. *See also* natural medicine
Wellness Real Estate, 132, 133
The Wellness Syndrome (Cederström), 138
Wells, H. G., 89
Wells, Ida B., 185
Weltmer, Sidney, 117–18
whales, 103–4
What Comes Naturally (Pascoe), 185–86
What to Expect When You're Expecting
    (Murkoff and Mazel), 18
wheatgrass, 108, 109
"Who Deserves to Be Healthy?" (Blon-
    diau), 143
Whole Foods Market, 110, 134–35, 136
Wigmore, Ann, 108, 109
wilderness, defining, 98–99
Wilderness Act (1964), 98
Wilkes, Samuel, 185
Williams, Joy, 103–4

Wills, Garry, 183–84
Wilmut, Ian, 11, 12
Wilson, Bee, 47
Wilson, E. O., 82, 221n22
wine, 45–46, 47
Winnie-the-Pooh, 90–91
The Wizard and the Prophet (Mann), 208
wolves, 94–99
women: agricultural labor by, 37;
    bodybuilding and, 192; displayed as
    biological specimens, 21; menstrual
    cycle of, 172, 176; sexism and sex
    segregation in sports, 6, 199–205. *See
    also* childbirth; natural family planning
    (NFP)
Wong, Debbie, 18
woolly mammoths, 206

Xidollien, 51
xocolatl, 47

Yakama, 102
Yan Fu, 166
Yellow Emperor's Classic of Medicine, 112–13
Yellowstone National Park: bison in, 99,
    100–103; camping trips in, 87–90,
    92; declaration of, 87, 101; humans
    in, 98–100; wolves in, 94–99. *See also*
    national parks
You Are the Universe (Chopra), 131
Young, Robert, 195

Zhuangzi, 66, 67
ziran, 3–4, 66
zoos, 87, 96